# Molecular Biophysics
## of the
## Extracellular Matrix

# Molecular Biology and Biophysics

**Molecular Biophysics of the Extracellular Matrix,** edited by *Struther Arnott, D. A. Rees,* and *E. R. Morris,* 1983

# MOLECULAR BIOPHYSICS OF THE EXTRACELLULAR MATRIX

Edited by

## STRUTHER ARNOTT

*Purdue University, West Lafayette, Indiana*

and

## D. A. REES and E. R. MORRIS

*National Institute for Medical Research, London, England*

Humana Press • Clifton, New Jersey

**Library of Congress Cataloging in Publication Data**

Main entry under title

Molecular biophysics of the extracellular matrix.

  (Molecular biology and biophysics)
  Bibliography: p.
  Includes index.
  1. Extracellular space.   2. Ground substance (Anatomy)
3. Molecular biology.   4. Biophysics.   I. Arnott,
Struther.   II. Rees, David Allan.   III. Morris, E. R.
IV. Series.
QH603.E93M65   1983        574.8        84-6640
ISBN 0-89603-051-2

©1984 The Humana Press Inc.
Crescent Manor
PO Box 2148
Clifton, NJ 07015

Printed in the United States of America

# Preface

Biopolymers, such as proteins and nucleic acids have been subjected to a rational succession of structural studies in which the determination of chemical compositions, linkages, and sequences is followed by investigations of molecular architecture and interactions. By the end of the 1960s the covalent structures of glycosaminoglycans and proteoglycans had been defined both within the carbohydrate chains and in respect to their linkages to protein. The scene was therefore set for successful probing of three-dimensional shapes and intermolecular associations. Fortunately, during the 1970s appropriate physical methods were themselves greatly increasing in power. Consequently much progress has been made using X-ray diffraction analysis of hydrated films, nuclear magnetic resonance spectroscopy, and chromatographic methods for investigation of molecular interactions in solution. We now have a great deal of information about molecular shapes, their sensitivity to environment (especially associated cations), and their modes of interaction that could be relevant to supermolecular assemblies in biological milieux. For these reasons we judged that it would be timely to distil the main conclusions from this phase of research to prepare for the next, which will involve detailed study of the interplay between biological function and molecular structure.

The idea for this volume emerged from a workshop meeting at Colworth sponsored by the Biochemical Society. Thereafter, we decided to organize a coherent, critical survey of the field, a book that would go beyond the sort of transcript of presentations that generally results from such meetings, and would take into account the spectrum of issues associated with research on the extracellular matrix. This volume is therefore an original treatise rather than a proceedings, a thoroughgoing compendium and digest of significant new physical insights that we hope will provide a strong foundation for studies of the molecular biology of the extracellular matrix in years to come.

*Struther Arnott*
*D. A. Rees*
*E. R. Morris*

## Acknowledgment

We wish to acknowledge with thanks the expert assistance of Mrs. Lesley Linger throughout the course of preparation of this volume for publication.

# Contributors

STRUTHER ARNOTT • *Department of Biological Sciences, Purdue University, West Lafayette, Indiana, USA*

B. CASU • *Istituto di Chimica e Biochimica "G. Ronzoni," Milan, Italy*

WAYNE D. COMPER • *Department of Biochemistry, Monash University, Clayton, Victoria, Australia*

L. CÖSTER • *Department of Physiological Chemistry, University of Lund, Lund, Sweden*

L.-Å. FRANSSON • *Department of Physiological Chemistry, University of Lund, Lund, Sweden*

T. E. HARDINGHAM • *Kennedy Institute of Rheumatology, Bute Gardens, Hammersmith, London, England*

TORVARD C. LAURENT • *Department of Medical and Physiological Chemistry, University of Uppsala, Uppsala, Sweden*

ALOK K. MITRA • *Department of Biological Sciences, Purdue University, West Lafayette, Indiana, USA*

E. R. MORRIS • *National Institute for Medical Research, Mill Hill, London, UK*

I. A. NIEDUSZYNSKI • *Department of Biological Sciences, University of Lancaster, Lancaster, UK*

C. F. PHELPS • *Department of Biological Sciences, University of Lancaster, Lancaster, UK*

BARRY N. PRESTON • *Department of Biochemistry, Monash University, Clayton, Victoria, Australia*

D. A. REES • *National Institute for Medical Research, Mill Hill, London, UK*

J. K. SHEEHAN • *Department of Biological Sciences, University of Lancaster, Lancaster, UK*

# CONTENTS

**CHAPTER 3**

# X-RAY DIFFRACTION ANALYSES OF GLYCOSAMINOGLYCANS

**Struther Arnott and A. K. Mitra**

**CHAPTER 4**

# CONFORMATION OF INDIVIDUAL RESIDUES AND CHAIN SEGMENTS OF GLYCOSAMINOGLYCANS IN SOLUTION BY SPECTROSCOPIC METHODS

**B. Casu**

## CHAPTER 5

# SELF-ASSOCIATION OF COPOLYMERIC GLYCOSAMINOGLYCANS (PROTEOGLYCANS)

L.-Å. Fransson, L. Cöster, I. A. Nieduszynski,
C. F. Phelps, and J. K. Sheehan

## CHAPTER 6

# TRANSPORT OF MOLECULES IN CONNECTIVE TISSUE POLYSACCHARIDE SOLUTIONS

Barry N. Preston, Torvard C. Laurent, and
Wayne D. Comper

## CHAPTER 7

# GLYCOSAMINOGLYCAN CHAINS IN THE BIOLOGICAL STATE
### D. A. Rees

# Chapter 1

# Structure and Associations of Proteoglycans in Cartilage

## T. E. Hardingham

Kennedy Institute of Rheumatology, Bute Gardens,
Hammersmith, London, UK

### Introduction

Proteoglycans are complex macromolecules in which many long carbohydrate chains are linked to a protein backbone. The main carbohydrate chains form a family of related structures, the glycosaminoglycans. These are relatively long, unbranched polysaccharide chains composed of repeating disaccharide units. Each disaccharide carries sulfate and/or carboxyl residues. Seven basic glycosaminoglycan types (Table 1) occur in mammalian tissues (see Muir and Hardingham, 1975). With the exception of hyaluronate, they have all been reported to occur linked to protein. The linkage to protein involves a neutral trisaccharide (gal-gal-xyl) at the reducing end of the glycosaminoglycan chains, with the xylose forming an O-glycosidic linkage with a serine residue of the protein. Keratan sulfate is linked differently. In skeletal keratan sulfate it is linked via an O-glycosidic linkage of N-acetylgalactosamine to serine or threonine. In corneal keratan sulfate it is linked via an N-glycosylamine linkage from N-acetylgalactosamine to as-

TABLE 1
Composition of Mammalian Glycosaminoglycans

| | Disaccharide repeating unit | | | |
| | Hexuronic acid | Hexosamine[a] | Sulfate | Linkage to protein |
|---|---|---|---|---|
| Hyaluronic acid | D-glucuronic acid | D-glucosamine | — | ? |
| Chondroitin 4-sulfate | D-glucuronic acid | D-galactosamine | O-sulfate | gal-gal-xyl-ser |
| Chondroitin 6-sulfate | D-glucuronic acid | D-galactosamine | O-sulfate | gal-gal-xyl-ser |
| Dermatan sulfate | L-iduronic acid or D-glucuronic acid | D-galactosamine | O-sulfate | gal-gal-xyl-ser |
| Keratan sulfate | D-galactose | D-glucosamine | O-sulfate | I: skeletal (NANA-gal)-galNAc(-ser or -thr) II: corneal galNAc-asn |
| Heparan sulfate | D-glucuronic acid or L-iduronic acid | D-glucosamine | O-sulfate and N-sulfate | gal-gal-xyl-ser |
| Heparin | D-glucuronic acid or L-iduronic acid | D-glucosamine | O-sulfate and N-sulfate | gal-gal-xyl-ser |

[a]Always N-acetylated except when N-sulfated.

paragine. Proteoglycans are thus complex glycoproteins and the information available concerning their biosynthesis suggests that they share the same pathways as other secreted glycoproteins.

There is a well-defined technology for isolating and characterizing the glycosaminoglycan chains (Rodén et al., 1972), but complete evaluation of the structure, distribution, and function of the different proteoglycans has yet to be carried out.

Typically proteoglycans are components of extracellular spaces. They are hydrophilic and their branched "bottle brush" structures and high charge densities enable them to occupy large domains from which other macromolecules are excluded, but which are freely permeable to small molecules (Comper and Laurent, 1978). They are thus important in creating intercellular fluid compartments.

Where proteoglycans are abundant, the intercellular compartment is large. In connective tissue it is developed to form a matrix in association with networks of fibrous components, collagen, and elastin. Tissues such as skin and blood vessel walls are "tough" and their properties can be related to their composition and the blend of fibrous and nonfibrous components. The fibrous elements provide the tensile properties and the overall shape of the tissue, whereas the proteoglycans extend the fibrous network by drawing water in and give the tissues resilience.

Cartilage is a tissue with a specialized biomechanical function. In it the extracellular matrix is particularly large and is composed of a dense meshwork of fine collagen fibers (Type II) embedded in a highly concentrated solution of aggregated proteoglycan molecules. The composition of hyaline cartilage is 1–10% cells (Hammerman and Schubert, 1962) and 75% water, 40–70% of the dry weight is collagen and 10–20% glycosaminoglycan (Stockwell, 1979).

The proteoglycans found in cartilage are of high molecular weight in the range $0.5 \times 10^6$ to $4.0 \times 10^6$ (Hascall and Sajdera, 1970). They contain many chondroitin 4- or 6-sulfate chains, fewer keratan sulfate chains, and some short oligosaccharides attached to a protein core of about $2 \times 10^5$ molecular weight. The composition of a typical hyaline cartilage proteoglycan is shown in Table 2.

They are not only of high molecular weight, but also form large aggregates in which many proteoglycans bind to

TABLE 2
Cartilage Proteoglycans[a]
Molecular weight: $1 \times 10^6 - 4 \times 10^6$; $\bar{M}_w$ $2.5 \times 10^6$

| Composition | No. of chains | Molecular weight | Total weight, % |
|---|---|---|---|
| Protein | 1 | 200,000–350,000 | 7–12 |
| Chondroitin sulfate | 100 | 20,000 | 85 |
| Keratan sulfate | 50 | 5500 | 7 |
| Oligosaccharides | 30 | 1200–2000 | 1–2 |

[a]Charged groups: Sulfate:      4500/mol.
                 Carboxylate: 4200/mol.
                 Total:        8700/mol.

a chain of hyaluronate. The native aggregate also contains a specific link protein that further stabilizes the proteoglycan–hyaluronate bond. Calculation shows that up to 200 proteoglycans can bind to a single hyaluronate chain of molecular weight $1.6 \times 10^6$, producing an aggregate 4 μm long and of particle weight $350 \times 10^6$ (Hardingham et al., 1981).

This aggregating-type of proteoglycan appears to be a specific component of the cartilage matrix. Chick embryonic mesenchymal cells do not produce the cartilage type of proteoglycans prior to differentiation into chondrocytes (De Luca et al., 1977). Together with Type II collagen, cartilage proteoglycans have been taken as a marker for chondrogenic expression. Although aggregating proteoglycans of the "cartilage-type" are characteristically produced by chondrocytes, they may also be expressed by some other cell types since proteoglycans from the lamina intima of aorta cross-react with antibodies raised against cartilage proteoglycans (Gardell et al., 1980) and glial cells in culture also produce proteoglycans that aggregate with hyaluronate (Norling et al., 1978).

## Structure

The primary structure of cartilage proteoglycans has not yet been entirely resolved and current work assumes a model of structure that remains open to further revision. The slowness to elucidate even the basic structure of cartilage proteoglycan can largely be attributed to the general features of proteoglycans. The protein accounts for only a small proportion of the molecular weight (7–20%) (Table 2) and the high carbohydrate density (one chain per four amino acid residues over the greater part of the molecule) creates major problems in characterizing the protein because it limits the application of normal amino acid sequencing techniques and has made more difficult the isolation of specific cleavage products. Because of their expanded structure in solution and very high molecular weight, proteoglycans behave non-ideally even in quite dilute solutions (Comper and Laurent, 1978). This makes the interpretation of their molecular weight and shape from their physical properties much more difficult. This difficulty is also compounded by the proteoglycans being both polydisperse and heterogeneous. Each preparation is a family of closely related molecules rather than a single molecular species.

The current model takes into account many of these factors (Fig. 1). The protein backbone is composed of a globular region with intramolecular disulfide bridges, is of low carbohydrate content, and forms a specific site for binding to hyaluronate and to link protein (see below). An adjacent region of extended polypeptide has a high proportion of the keratan sulfate chains attached to it. The largest portion is a further extended polypeptide that contains the majority of the chondroitin sulfate chains (see Hascall, 1977). The chondroitin sulfate attachment region is proposed to be of variable length (Fig. 1), and this accounts for the polydispersity and the changing composition within each proteoglycan preparation, which shows the largest molecules to be of highest chondroitin sulfate content and lowest protein content. Electron micrographs of proteoglycans spread in monolayers shows the protein core of proteoglycans attached to aggregates to be of variable length (Rosenberg et al., 1975; Kimura et al., 1978). Further evidence for this structural model has also been obtained by specific enzymatic "dissection" of the molecule. Digestion of proteoglycan aggregates with chondroitinase ABC and trypsin (Heinegård and Hascall, 1974) yields a hyaluronate binding region fragment of 60,000–80,000

Proteoglycan                Proteoglycan aggregate

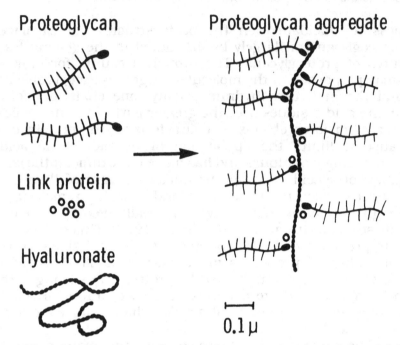

Link protein

Hyaluronate

0.1 μ

Fig. 1. Structure of proteoglycan aggregate (total weight $20 \times 10^6$ to $500 \times 10^6$).

molecular weight, and a keratan sulfate-rich peptide of about 30,000 molecular weight (Hascall, 1977). Recent evidence has shown that there are also many smaller oligosaccharides attached to the proteoglycan. These are both O-glycosidically-linked oligosaccharides that occur mainly in the keratan sulfate-rich region and N-glycosylamine-linked oligosaccharides that are on the hyaluronate binding region (Thonar and Sweet, 1979; De Luca et al., 1980; Lohmander et al., 1980).

Although the present model of proteoglycan structure provides a rational framework that accounts for many features, it cannot be taken as complete or final. For example, the separation of proteoglycan monomer preparations from a variety of cartilaginous sources into two or three discrete components on electrophoresis (Stanescu et al., 1973, 1977; Roughley and Mason, 1976) is not predicted from the model and may indicate that further elaboration is needed.

A major feature of the model is that it permits the interpretation of some of the variations in proteoglycan size and composition amongst different cartilaginous tissues, and the changes that occur with aging, on the basis of a single composite structure (Figs. 2 and 3). The proteoglycans of different cartilages may be based on polypeptides to which variable numbers and sizes of glycosaminoglycan chains and oligosaccharides have been attached (Stevens et al., 1979). For example, some of the shorter O-glycosidically linked oligosaccharides may reflect abortive keratan sulfate chains and may be the sites at which full length keratan sulfate chains are synthesized on the proteoglycans of older tissues. According to this model, the major polydispersity

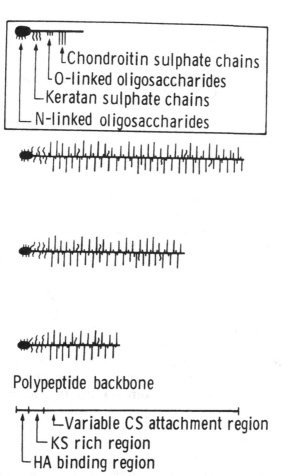

Fig. 2. Cartilage proteoglycan: model of structure and polydispersity.

a)

b)

c)

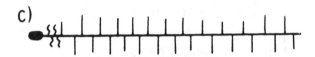

d)

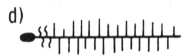

Fig. 3.   Variation in structure of cartilage proteoglycans: (a) standard model; (b) shorter chondroitin sulfate chains; (c) fewer chondroitin sulfate chains; (d) shorter chondroitin sulfate region.

within any one preparation arises from the variable length of the chondroitin sulfate attachment region. The origin of this variation in length has not been established, but since newly synthesized proteoglycans are suggested to be less polydisperse than those extracted from the tissue (Oike et al., 1980; Fellini et al., 1981), it is likely that some of the polydispersity may result from degradative action in the extracellular space. Within a single proteoglycan preparation there are reports of both aggregating and nonaggregating proteoglycans, but it remains to be determined whether these represent proteoglycans based on different polypeptides, or variations in the secondary modifications of a single polypeptide product.

## Preparation

The advances in our understanding of structure owe much to the development of efficient, nondegradative methods for extracting and purifying proteoglycans. This has helped resolve the conflicting evidence of structure, size, and polydispersity that arose when methods of extraction gave variable yields, and a wide variety of fractionation methods led to the isolation of different samples of the proteoglycan population. Proteoglycans can also be degraded by high shear forces (Sajdera and Hascall, 1969a) and by proteolysis during extraction (Oegema et al., 1975). This can be difficult to detect, since purified proteoglycans appear to be both polydisperse in size and heterogeneous in composition, but the use of low shear extraction procedures and inhibitors of proteolysis have greatly reduced the risk of degradation from these sources.

The work of Sajdera and Hascall (1969a) established that a high proportion of proteoglycans could be extracted from cartilage in $4M$ guanidinium chloride (GuCl) without the use of high speed homogenization or other disruptive treatments. They also developed the use of equilibrium density gradient centifugation in cesium chloride (CsCl) solutions for the purification of proteoglycans. This had the benefit of being a mild procedure of high capacity that permitted the purification of proteoglycans of high buoyant density without bringing them out of solution at any stage. They showed that proteoglycans reversibly aggregated. Since they were dissociated in $4M$ GuCl, but remained aggregated in concentrated CsCl (Hascall and Sajdera, 1969), it was possible to isolate aggregated proteoglycan from a CsCl gradient with low GuCl concentration ($0.5M$) or disaggregated proteoglycan from a gradient with high concentration of GuCl ($4.0M$) (Fig. 4). Aggregates were shown to be of high sedimentation coefficient in the ultracentrifuge (50–60 S) compared with a monomer sedimentation coefficient of 24 S. Aggregate and monomer were also shown to be separated by gel chromatography on Sepharose 2B. In the CsCl gradient in the presence of $4M$ GuCl, not only are the proteoglycans largely separated from hyaluronate and link protein, but there is also some fractionation of the proteoglycan population such that molecules of higher pro-

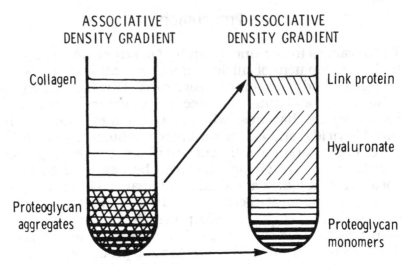

Fig. 4. Left: Associative density gradient: fresh sliced cartilage extracted with 4$M$ GuCl buffered at pH 5.8 or 4.5; extract dia-
lyzed to 0.5$M$ GuCl, pH 5.8; CsCl added to give a density of
1.50–1.69 g mL$^{-1}$; centrifuged at 100,000 $g_{av}$ for 48 h at 20°C.
Right: Dissociative density gradient: the lower fraction from the
associative gradient mixed with an equal volume of 7.5$M$ GuCl
buffered at pH 5.8 and the density adjusted with CsCl to 1.50 g
mL$^{-1}$; centrifuged at 100,000 $g_{av}$ for 48 h at 20°C.

tein content occur in fractions of lower buoyant density.
Analysis of the proteoglycan fractions of different buoyant
density revealed a change in composition and in size
(Tsiganos et al., 1971).

Investigation of the components of aggregation led to
the identification of hyaluronate among the density gradient fractions (Hardingham and Muir, 1973a). It accounted
for about 1% of the glycosaminoglycan in cartilage and was
shown to bind to proteoglycan. This interaction involving
two large polyanions was unexpected, but was shown to be
highly specific, and the stoichiometry suggested that many
proteoglycans were binding to each hyaluronate chain
(Hardingham and Muir, 1972b). In addition to hyaluronate,
a glycoprotein component was released from the aggregate
in the density gradient. This link-protein component was
shown to bind to the aggregate and stabilize the proteoglycan—hyaluronate bond as described below.

## Link-Free Aggregates

The binding of proteoglycan to form link-free aggregates has been characterized in some detail (Hardingham and Muir, 1972b, 1974; Hascall and Heinegård, 1974; Christner et al., 1977, 1979; Nieduszynski et al., 1980). Aggregates rapidly form when proteoglycan and hyaluronate solutions are mixed in physiological saline or many other buffers at neutral pH. The aggregate formed is completely dissociated at acid pH and in high concentrations of GuCl, $CaCl_2$, $MgCl_2$, or in $6M$ urea or 0.1% (w/v) sodium dodecyl sulfate (SDS), but not in high concentrations of NaCl or CsCl. It is also reversibly dissociated on heating to 60°C. The binding site is only slowly and irreversibly denatured at higher temperatures ($t_{1/2}$ = 140 min at 80°C), but binding of proteoglycan to hyaluronate is abolished by several chemical modifications of amino acids that do not involve polypeptide cleavage. The acetylation of lysine and other amino groups, the modification of arginine with butanedione, the oxidation of tryptophan with $N$-bromosuccinimide and its reaction with 2-nitrophenyl sulfenylchloride all abolished binding. The substitution of lysine residues by reaction with 2-methyl maleic anhydride or the reduction of disulfide bridges also prevented binding. After these treatments binding could be largely restored by removal of the methyl maleyl groups at pH 3, and by the reoxidation of disulfide bridges. The binding region of proteoglycan is thus remarkably stable thermally, but is easily perturbed by alterations in the structure of several amino acids, which therefore appear to be necessary for maintaining an active binding conformation (Hardingham et al., 1976).

The binding site is very specific for hyaluronate and has a high affinity for decasaccharide ($HA_{10}$) or larger fragments ($K_D$, $3 \times 10^{-8}M$) (Nieduszynski et al., 1980), but lower affinity for $HA_8$ or smaller fragments (Hardingham and Muir, 1973b; Hascall and Heinegård, 1974). Chemical modification of hyaluronate suggest that at least four carboxyl groups on a decasaccharide unit are necessary for strong binding, since it was greatly decreased by other bulky substituents in their place or by their reduction to the alcohol (Christner et al., 1977, 1979). The lack of binding of desulfated chondroitin sulfate (chondroitin) suggests considerable specificity on the hexosamine moiety since this differs from hyaluronate only at the hydroxyl position at C(4) (Hascall and Heinegård, 1974).

## Link-Stable Aggregates

The nature of the interaction of proteoglycan with link protein has not been characterized in detail, and much of the evidence of the way in which link protein functions has come from more indirect observations (Hardingham, 1979; Tang et al., 1979; Kimata et al., 1982). In the absence of link protein, proteoglycans bind reversibly to hyaluronate, and this is open to competition by oligosaccharides—but with link-stable aggregates there is no longer competitive binding with oligosaccharides, and no dissociation is evident (Hardingham, 1979) (Fig. 5). The attachment of proteoglycan to hyaluronate is therefore much stronger, and the link protein effectively locks proteoglycan onto the hyaluronate chain. In vitro the stabilizing effect of link protein is evident on heating up to 50°C in $0.5M$ guanidinium chloride and even up to 60°C in $0.15M$ NaCl buffered at pH 7.4 (Hardingham, 1979). The link-stable aggregate is dissociated similarly to the link-free aggregate in high concentrations of GuCl, $CaCl_2$, $MgCl_2$, and SDS, but it is not dissociated in $4M$ urea. Even when proteoglycans are dissociated from hyaluronate, however, there is some residual binding of link protein to proteoglycan (Kimata et al., 1982).

Link protein has been shown to exist in two molecular weight forms in several cartilaginous tissues. Evidence suggests that the two forms are structurally related, with the smaller form lacking a glycopeptide present on the larger form (Baker and Caterson, 1979). Both forms appear to be functionally active, and there is 1 mol of link/mol of proteoglycan (Kimura et al., 1980).

Link protein can be prepared from proteoglycan aggregates by isolating the fraction resistant to trypsin digestion that contains both the hyaluronate binding region and the link protein (Heinegård and Hascall, 1974). In this preparation the link is entirely of low molecular weight and may be even smaller than the naturally occurring small form (Baker and Caterson, 1979), but it still retains its function. The

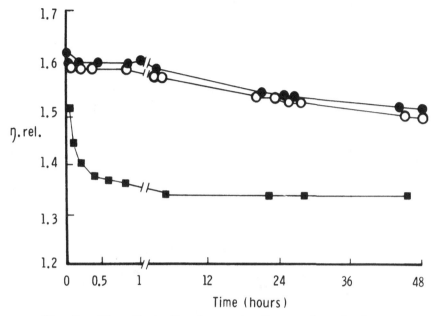

Fig. 5. The effect of hyaluronate oligosaccharides (accounting for 2% of total uronic acid) on the viscosity of proteoglycan–hyaluronate aggregates (0.5 mg mL$^{-1}$; pH 5.8; 30°C; 0.5$M$ GuCl). Results are shown for link-stabilized aggregate alone (●) and after the addition of HA$_{38av}$ (○) and for link-free aggregate after addition of HA$_{38av}$ (■). Closely similar behavior was observed with HA$_{10}$ oligomers. (Hardingham, 1979).

significance of the two forms is therefore not clear, and although it appears likely that the smaller may be derived from the latter by proteolytic cleavage, examination of newly synthesized and secreted radioactive link protein in cultures of chondrosarcoma cells showed only the smaller form, which suggests that if any selective cleavage occurs, it must be intracellular and prior to secretion (Kimura et al., 1980).

## The Secretion and Assembly of Proteoglycan Aggregates

Proteoglycans are synthesized by chondrocytes and secreted into the extracellular matrix. Autoradiography of sections of cartilage pulse-labeled with $^{35}$S-sulfate suggests that movement of proteoglycans out of chondrocytes and

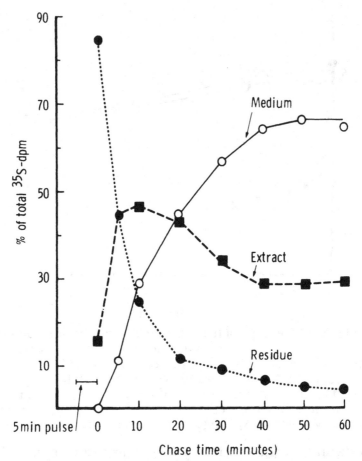

Fig. 6.   Pulse-chase of $^{35}$S-sulfate on chondrocytes in culture: cell residue fraction (●); 4$M$ GuCl extract (■); culture medium (○) (Kimura et al. 1979).

into the matrix is fairly rapid (Hardingham and Muir, 1972a).

The assembly of aggregates appears to provide a mechanism for immobilizing proteoglycans within the collagenous matrix of cartilage. The binding of link-protein provides a potentially irreversible step in aggregate formation. It is therefore important to determine whether aggregation occurs intracellularly or after secretion of the proteoglycans into the cartilage matrix. Evidence for the extracellular assembly of proteoglycan aggregates was obtained with cultures of chondrocytes from a chondrosarcoma (Kimura et al., 1979). Examination of the proteoglycans secreted into the medium in pulse-chase experiments showed that at early times (10–20 min) mainly proteoglycan monomer was present, but they were steadily incorporated into aggregates over 2 h. By adding an excess of cold carrier proteoglycan to the radioactive sample it was possible to distinguish between link-free and link-stable aggregates, and it was therefore possible to follow link-stabilization in the medium (Figs. 6 and 7). Oligosaccharides of hyaluronate derived by digestion with testicular hyaluronidase were also used as probes of stable aggregate formation. $HA_{16}$ oligosaccharides were found to delay stable aggregate formation in the medium and to increase the flow of proteoglycan from the pericellular matrix into the medium. With larger oligosaccharides ($HA_{50}$) that were able to bind to both proteoglycan and link-protein simultaneously, the formation of stable aggregates was blocked completely (Fig. 8). The results showed that the formation of link-stable aggregate was an extracellular event since the oligosaccharides were only able to compete with hyaluronate in binding to proteoglycan prior to link-stabilization. The oligosaccharides were also only effective if present at a time when proteoglycans were newly secreted from the cells. Although the cells used for this study were from a chondrosarcoma, similar displacement effects with oligosaccharides were also observed with normal chick chondrocytes, suggesting that proteoglycans are also secreted prior to link-stabilization in this system (Solursh et al., 1980). These relationships are summarized in Fig. 9.

Although the results indicate that aggregate stabilization occurred only after secretion, it is not clear what mechanism delays it. All three components of aggregates are synthesized by the chondrocytes, and if they are present within the same compartment within the cell, why should they not interact? Experiments using exogenous proteoglycans suggest that link-protein and proteoglycan may be bound together prior to secretion (Kimura et al., 1980). The limiting factor in aggregate formation might therefore be the availability of hyaluronate. The site of synthesis of hyaluronate in the cell is not known, and it may well be separate from the normal routes of synthesis of secretory proteins. This suggestion is supported by studies with monensin that showed

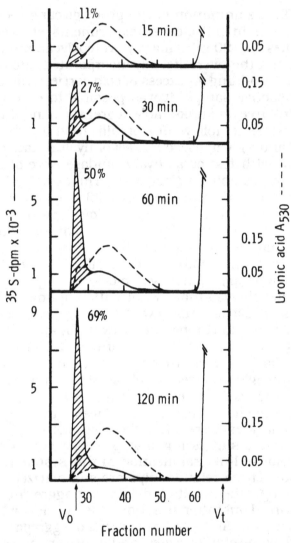

Fig. 7. Formation of link-stabilized proteoglycan aggregate in the medium of chondrocyte cultures at various times after pulse-labeling with $^{35}$S-sulfate (Kimura et al., 1979).

it to have no effect on hyaluronate synthesis in chondrocytes at concentrations that potently inhibited chondroitin sulfate synthesis on proteoglycan (Mitchell and Hardingham, 1982). Furthermore, other mechanisms involving the conversion of link-protein or proteoglycan

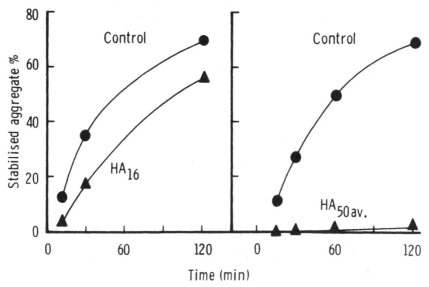

Fig. 8. Effect of hyaluronate oligosaccharides, $HA_{16}$ and $HA_{50av}$, on the formation of link-stabilized aggregate in the medium of chondrocyte cultures. Plates were pulsed for 5 min with $^{35}S$-sulfate, and the appearance in the medium of stabilized aggregate was determined at subsequent times by gel chromatography, as described by Kimura et al., (1979).

from inactive to active binding forms during intracellular processing cannot be ruled out. In the chondrosarcoma cell system, link-protein appears to be synthesized in excess over the amount of proteoglycan (Kimura et al., 1980). It remains to be determined what coordination there is in the controls of synthesis of the components of aggregate.

The role of proteoglycans in cartilage is dependent on them providing a very high concentration of fixed negative charge in the tissue. Aggregation can be seen as a way of immobilizing the proteoglycans within the network of collagen fibers. The extracellular assembly of aggregates and the stabilization with link-protein core is thus analogous to the secretion and assembly of collagen fibers although the mechanisms involved are entirely different. The origin and significance of the large variation in cartilage proteoglycan structure and the polydispersity remain unclear, but it is evident that there is immense scope for modulation of the structure during the post-ribosomal stages of synthesis,

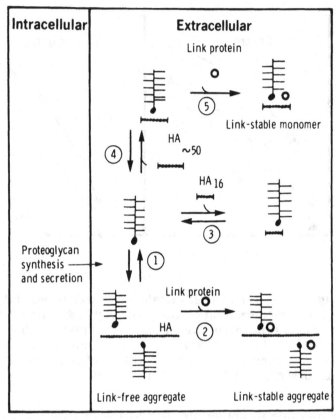

Fig. 9. Formation of link-stabilized aggregates in cell culture, and its inhibition by hyaluronate oligosaccharides. (1) Reversible binding of proteoglycan and hyaluronate. (2) Irreversible link-stabilization of aggregate. (3) Reversible binding of proteoglycan to $HA_{16}$. (4) Reversible binding of proteoglycan to $HA_{50av}$. (5) Irreversible link-stabilization of proteoglycan–$HA_{50av}$ complex.

and it remains an important challenge to evaluate how far this is correlated with the functional requirements of the tissue.

## Diversity of Proteoglycan Structure

Cartilage proteoglycans aggregate in a very specific way, and although they may not be exclusive to cartilage it is clear that in other tissues there are proteoglycans that do not aggregate in this way. For example dermatan sulfate proteoglycans are unable to bind to hyaluronate but appear to self-associate by interaction among the polysaccharide chains (Fransson et al., 1979). The survey of different proteoglycan types is far from complete, but there is sufficient evidence to show that the same glycosaminoglycan chains can also occur in different types of proteoglycans. The chondroitin sulfate proteoglycans produced by ovarian follicle cells (Yanagishita et al., 1979) are of quite distinct structure from cartilage proteoglycans, although of similar size, and they are very different from the smaller chondroitin sulfate proteoglycans produced by pre-chondrogenic cells (De Luca et al., 1977). The same carbohydrate chains are thus being synthesized in different cell types attached to different protein core primers.

Our understanding of the range of proteoglycan types and their morphological distribution thus depends on knowing the structure of their protein components. It will then become apparent how far the information that determines the structure of a proteoglycan resides in the protein core, or how far it is directed by the cell's complement of enzymes, which are involved in its post-translational processing. In understanding the physiological function of proteoglycans it is clear that the glycosaminoglycan chains provide the dominant physical properties, but the protein core may be of importance in influencing the presentation of the chains. There are a number of ways in which the protein core may influence the properties of the glycosaminoglycan chains. This includes how closely the glycosaminoglycan chains are packed along the protein core and whether this permits the formation of ordered arrays. It is also likely to be influenced by the number and type of oligosaccharides that are attached to the protein core. These may also have special properties of interaction of their own. The presence of ordered protein structures of specialized function (such as the hyaluronate-binding region) may also influence the properties. The protein core is thus not only important in establishing the relationship between different proteoglycan types, but it can have a directing influence itself on the properties of the glycosaminoglycan chains and how they interact with each other, with ions, water, and with other connective tissue components.

# Chapter 2

# The Dilute Solution Properties of Glycosaminoglycans and Proteoglycans

## C. F. Phelps

Department of Biological Sciences, University of Lancaster, Lancaster, UK

### Introduction

It may be a reaction to the present economic pressures that prompts the question "Have we obtained good value for our money in hydrodynamic studies of glycosaminoglycans and proteoglycans?" Sad as it is, my own feeling is, no. Confronted with the plethora of higher information content data derived using instrumentation such as NMR spectrometers or X-ray diffractometers, the contribution of those toilers at the mill where centrifuges, viscometers, and light scattering photometers abound has been small. Thus, in an arguably appropriate use of the word, it would be profitable to review what has been achieved. I shall choose to be eclectic since only in this way can some connectivity between parts be imposed on such wayward subject matter.

## The Relevance of the Term "Dilute Solution"

Apart from the academic interest of observing the properties of macromolecules in this quasi-ideal thermodynamic environment, it might be asked if these conditions in any way approach the real world of physiology. Table 1 summarizes some of the known concentration ranges in which the glycosaminoglycans and proteoglycans exist in some tissues. The data are based on percentage of the tissue wet weight, which is misleading. The needful term is the concentration found in the extracellular matrix and even this may need correcting for the insoluble fibrillar components of that compartment.

Whatever the modulating factors, it can be seen that the connective tissue polysaccharides are found primarily in "dilute solution" in the extracellular space. Significantly, the cartilage proteoglycans are not, and since most of our

TABLE 1
Overall Concentrations of Glycosaminoglycans in Some Tissues[a]

| Tissue | Glycosaminoglycan | Concentration, % tissue wet weight |
|---|---|---|
| Human synovial fluid | Hyaluronic acid | 0.30 |
| Human vitreous body | Hyaluronic acid | 0.014–0.034 |
| Bovine vitreous body (various parts) | Hyaluronic acid | 0.01–0.1 |
| Human umbilical cord | Hyaluronic acid | 0.8–1.0 |
| Cock's comb | Hyaluronic acid | 0.096 |
| Rabbit aortic wall | Not separated | 0.4 |
| Human skin | Not separated | 0.04–0.08 |
| Tadpole | | |
|     Back skin | 96% hyaluronic acid; 4% chondroitin sulfate | 0.10 |
|     Tail fin | 96% hyaluronic acid; 4% chondroitin sulfate | 0.13 |
| Cartilage (puppies) | Chondroitin sulfate | 5.5–8.0 |
| Bovine cornea | Keratan sulfate; chondroitin sulfate | ca. 1.0 |

[a]From Laurent et al. (1969).

information on proteoglycans comes from this family of molecules, we must be careful in interpreting properties found in dilute solutions as being of physiological significance.

## Hyaluronate

To what questions would we like answers? Chemical information consistently reinforces the belief that hyaluronate is an unbranched chain made up of invariably alternating units of two sugar derivatives. The first contribution of hydrodynamic studies was to show the size of this macromolecule (Blix and Snellman, 1945; Ogston and Stanier, 1951). It can be a very large molecule indeed, with weights reported in the millions. Very few molecules that are not template-directed in their biosynthesis achieve this size, and it should lead the enquiring mind to thoughts of error frequency and their implications. No information is presently available.

From this mundane zero-order statement we can structure further questions, among the next of which might be one related to the constraints on the molecule in solution. A long molecule in perfect thermodynamic agreement with its solvent would have those free energy terms relating to polymer–polymer, polymer–solvent, and solvent–solvent interactions almost equal to each other, and the expected conformation would then be that of a flexibly coiling molecule. The techniques used for establishing the solution conformation are many and varied, and most of them have the reassurance of being grounded in competent physical polymer theories.

The simplest formulation is that of Mark-Houwink (Houwink, 1940), where the dependence of the intrinsic viscosity of a linear chain molecule on its molecular weight is expressed by $[\eta] = KM^{\alpha}$. For flexible polymers the exponent $\alpha$ can have values in the range 0.5–0.8. As elements of stiffness enter into the conformation, the value of $\alpha$ rises above 0.8. Cleland and Wang (1970) analyzed a number of fractionated samples of vitreous human hyaluronate as well as rooster comb material. Their data are shown in Fig. 1, and three issues are raised. First, the exponent $\alpha$ of the Mark-

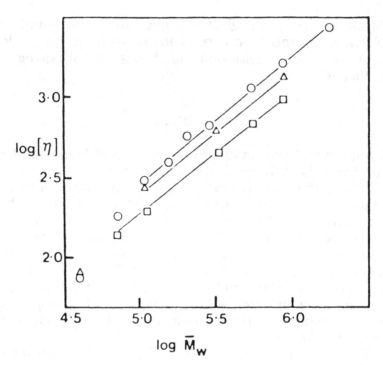

Fig. 1. Double logarithmic relation of $[\eta]$ to $\overline{M}_w$ for hyaluronate in 0.2M NaCl (o), 0.5M NaCl (Δ) and 0.1M HCl (□).

Houwink equation is 0.8 in 0.5M NaCl. This value is at the allowed limit for a freely coiling molecule. However, the polyelectrolyte nature is seen in the value of α = 0.816 in 0.2M NaCl. Lastly, and surprisingly, there is only a small fall in a numerical value of α (to 0.763) when the charge on hyaluronate is titrated out in 0.1M HCl.

Other high polymer systematics have also been used by the same authors. The dependence of the sedimentation coefficient on molecular weight can be expressed in a form where another exponent lies near the value for coiled structures (Kurata and Stockmayer, 1963), whereas the hybridized Mandelkern-Flory formulation (Mandelkern et al., 1952) should express, in the β function, a parameter independent of polymer–solvent systems. Flexible polymers often have values of β near (2.6–2.7) × $10^{-6}$ dL $g^{-1}$. Hyaluronate gives values of 2.9–3.1. In the exhaustive treatment of Cleland and Wang, the second virial coefficient

$(A_2)$, measured either by sedimentation and diffusion or by direct osmometry, varies between $2.2 \times 10^{-3}$ and $1.8 \times 10^{-3}$ mol cm$^3$ g$^{-2}$ for hyaluronate in $0.5M$ NaCl. The theoretical magnitude of $A_2$ calculated from considerations of the Donnan effect is $A_2 = 1000/4 M_o^2 m_3$, where $M_o$ is the equivalent weight to the polymer, and $m_3$ is the molar concentration of added electrolyte. In $0.5M$ NaCl, $A_2$ would then be $2.7 \times 10^{-3}$ mol cm$^3$ g$^{-2}$. The agreement between theory and experiment in $0.2M$ NaCl is, however, not so impressive.

What we have therefore is a suggestion that, under physiological conditions, hyaluronate has large domains of freely coiling structure, but that there are elements of stiffness in the molecule.

Can we produce an estimate of how much stiffness there is in these structures? The inexhaustable Cleland has attempted just that (Cleland, 1970a). Again using methods pioneered for high polymer chemistry, Cleland has calculated the root-mean-square end-to-end distance for a theoretical model of a pyranose polymer that possesses complete rotational freedom around each of the glycosidic bonds. This laxity is called into question by calculations (Rees, 1969) of steric restraint in the glycosaminoglycans. When Cleland's calculated value is compared with the experimentally measured unperturbed mean-square end-to-end distance, the discrepancies that are seen may be attributed to hindrance of rotation and thus give some quantitative measure of stiffness. The experimental values are derived either from extrapolated values of [$\eta$] in good solvents (Flory and Fox, 1951) or can be determined from small-angle X-ray scattering experiments (Cleland, 1977) which give information on the persistence length and cross-sectional radius of gyration of the polymer (Kratky, 1966).

The consolidated position arrived at is that hyaluronate is a "moderately" or "partially" stiffened coiled molecule. In a time when many complicated molecules are architecturally defined to a fraction of an Ångstrom unit, such "molecular" information may at first be thought to be vapid.

However, it remains in the memory how few times hyaluronate obeys the canon of the well-behaved high polymer. This dissembling may be at the root of the physiological function of hyaluronate and reflect the very small changes in free energy required for the molecule to change conformational postures. Thus the time-average measurements by hydrodynamic methods are probably not relevant (or good value) to any other attribute of the molecule than its weight, volume of occupancy, or its frictional attributes.

A number of hydrodynamic techniques have yielded disappointingly little information. The universally accessible technique of gel chromatography is a case in point. Despite the best efforts of Pharmacia, no commercially available gel matrix is yet capable of complete resolution of all the molecular weight classes present in any sample of hyaluronate. There is a perilous conceit in glycosaminoglycan and proteoglycan studies that approves of a matrix that excludes some portion of a sample while retarding other portions. For this reason little analytical, quantitative data is available from such work (Barker and Young, 1966). Use of ion exchange materials in the hands of Cleland (1970b) has revealed the distribution of molecular weights present in hyaluronate from a number of sources. A more direct way to the same end is revealed in the work of Fessler and Fessler (1966) who literally portray the distribution of lengths in an electron microscopic representation. There is some hope that porous glass beads may offer useful media for the examination of molecular weight profiles.

Many of the problems of extracting meaningful data from hydrodynamic studies are readily seen in the most obvious attribute of hyaluronate—its viscosity. This property shows a complicated dependency on extrinsic factors, such as concentration of hydrogen ion, concentration and type of salt, shear rate, and temperature. The first studies by Ogston and Stanier (1950–1952) suggested that the very high intrinsic viscosity could be best rationalized by considering hyaluronate as occupying a massively hydrated spherical domain in which, for a molecule of weight $8 \times 10^6$, 1000 times more water than polymer was contained. The shear-dependence of hyaluronate is shown in Fig. 2 and demonstrates a functional attribute of hyaluronate found in articular joints that will be apparent to anyone contemplating the lubrication needs of a human tiptoeing or sprinting.

A deeper analysis of the elastic and viscous components of viscosity was undertaken by Gibbs et al. (1968). At low strain-frequency, the viscous behavior is predominant, but

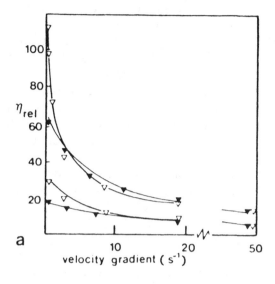

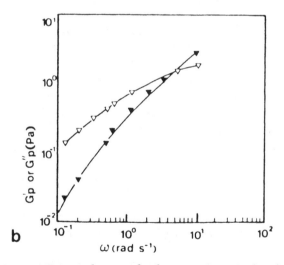

Fig. 2.    (a) Dependence of relative viscosity ($\eta_{rel}$) on velocity gradient at 0.5°C ($\nabla$) and 37°($\blacktriangledown$) for bovine synovial fluid hyaluronate at 0.8 mg mL$^{-1}$ (upper curve) and 0.5 mg mL$^{-1}$.
(b) Frequency dependence of storage modulus ($G'$;$\blacktriangledown$) and loss modulus ($G''$;$\nabla$) for rooster comb hyaluronate at 25°C, $I$ = 0.2, c = 3.0 mg mL$^{-1}$, pH = 7.0.

as the frequency increases, the elastic properties become
increasingly important. The "crossover point" at physiological concentrations of polymer lies near the requirements
experienced in walking. Increase in polymer concentration
also increases the elastic component. All these findings
gain credibility when seen in the context of an expanded
three-dimensional network of interpenetrating chains,
where contacts may be both of physical entanglement and
of chemical hydrogen bonding. Support for this view comes
from the detailed NMR studies of Welti et al. (1979), as well
as from the chemical studies of Scott and Tigwell (1978).
Two extreme pH values provoke opposite behavior in the
polymer. At pH 2.5 a remarkably sharp transition into, and,
at fractionally lower pH, out of, a viscoelastic putty is observed, possibly because of an optimization of interchain
hydrogen bonding, whereas above pH 11.0, the viscosity of
hyaluronate drops, allegedly by hydroxyl-group ionization.

Shear-dependent studies include those experiments on
streaming birefringence, the most recent of which are reported by Barrett (Barrett and Harrington, 1979; Barrett,
1978).

Sedimentation velocity and diffusion are transport
methods in which the frictional coefficient becomes a dominant restrictor of analytical information. The sedimentation coefficients are strongly dependent on concentration
and the way in which extrapolation to infinite solution is
made is still a matter of some dispute (see Preston et al.,
1965). The three-dimensional network offers hydraulic con-
straint to the sedimentation process, and indeed Ogston
has shown that

$$S/(1-\bar{v}\rho)c$$

is formally a hydraulic conductivity useful in measurements of physiological fluid exchange across extracellular
territories.

Diffusion processes will be similarly affected (Laurent
et al., 1960). Some elegant studies on model connective tissue systems were initiated by Ogston and Sherman (1961)
and have developed into the concentration solution studies
described by Preston et al. (this volume).

Whereas the transport processes are dominated by the
magnitude of the frictional coefficient, equilibrium meas-

urements such as those reported by Preston and Wik (1979) show what can be derived from careful experimentation.

In sedimentation equilibrium, using customary symbolism

$$\omega^2 c(1-\bar{v}\rho)/2RT(dc/dr^2) = (1/M) + 2A_2c + 3A_3c^2 + 4A_4c^3$$

The virial coefficients reflect the inappropriateness of the solvent environment, and a great deal of information is entombed in these constants.

Provided $A_2$ is large in comparison to $1/M$, the right-hand side of the equation reflects nonideality. According to Flory (1953), $A_2$ can be related to the effective diameter of the macromolecule and to an excluded volume term. As such it will not be a constant at different ionic strengths and the values derived by computer fitting of a polynomial are shown in Table 2. It is noteworthy that $A_4$ can adopt negative values, particularly at very low ionic strengths, as if it were compensating for the changes in $A_2$ with decreasing ionic strength.

So, hyaluronate at extreme dilution behaves in a manner for which our best model is a free coiling molecule with some restraints on total freedom of rotation. Because of the vast volumes of occupancy (more than 10 dL $g^{-1}$), at concentrations above 1 mg mL$^{-1}$ chain entanglement is seen, with extensive network penetration producing the higher-

TABLE 2
Viral Coefficients of Hyaluronate Determined by High-Speed Sedimentation Equilibrium

| Solvent NaCl, $M$ | $A_2 \times 10^3$, mol cm$^3$ g$^{-2}$ | Effective diameter, nm | $A_3 \times 10^2$, mol cm$^6$ g$^{-3}$ | $A_4$, mol cm$^9$ g$^{-4}$ |
|---|---|---|---|---|
| 0.2 | 4.1 ± 0.61 | 26 | 6.3 ± 3.0 | — |
| 0.05 | 8.6 ± 2.7 | 33 | 42 ± 30 | — |
| 0.01 | 37 ± 15 | 54 | 250 ± 150 | −47 |
| 0.005 | 43 ± 11 | 57 | 2400 ± 110 | −3,300 ± 1,200 |
| 0.001 | 170 ± 100 | 90 | 20,000 ± 13,000 | −81,000 ± 62,000 |

order viscous effects discussed in the text. The extent and permanence of the contacts are the subject of much current study.

## The Sulfated Glycosaminoglycans

Heparin is the only member of this family existing as a free polysaccharide chain, being cleaved off a macromolecular precursor by endoglycosidase action. All other members are found as subpendant chains to a protein core in proteoglycans, from which they may be removed by glycosidase or alkali action [see Table 3, which gives details of the molecular features to be observed in this group of polymers from the excellent review of Comper and Laurent (1978)].

There seems little point in arguing the case for any specific solution conformation from the scanty data available. Two points are apparent. The charge density can rise to the extraordinary level seen in heparin and the importation of such bulky charges in the form of sulfate ester groups must both restrict the free rotation around the bonds of the glycosidic bridges to an extent extra to that seen in hyaluronate as well as make the polyelectrolyte character of the polymer more pronounced.

Thus for chondroitin 4-sulfate, the Mark-Houwink exponent often exceeds 1, and the sedimentation coefficients (now accessible to experimental measurement) are proportional to the square root of the molecular weight, with a nonzero intercept as required by the theory of Hearst and Stockmayer (1964) [see Fig. 3, derived from data of Wasteson (1971)]. Though there are assumptions to be eschewed, chiefly that θ solvent conditions obtain, there is little to cloud the view that we are dealing with a close cousin of the free coil, whether it be a wormlike chain or a broken rod. Theoretical calculations marry well with the dimensions derived from solid-state structural studies.

Heparin is a special case, however, and deserves some treatment. The extensive work of Stivala and his group [see Stivala and Ehrlich (1974); Ehrlich and Stivala (1974)] have used the whole panoply of hydrodynamic tools and have done their best to fit this data to a bewildering number of

TABLE 3

Molecular Features and Distribution of Glycosaminoglycans of Vertebrate Connective Tissue[a]

| Polysaccharide | Molecular weight range, $\times 10$ | Charge per disaccharide unit[b] | Site of occurrence, examples |
|---|---|---|---|
| Hyaluronate | 4000–8000 | 1.0 | Ubiquitous in connective tissues (?), skin, synovial fluid, vitreous humor, heart valve, cartilage |
| Chondroitin 4-sulfate | 5–50 | 1.1–2.0 | Cartilage, bone, cornea, notochord, skin |
| Chondroitin 6-sulfate | 5–50 | 1.2–2.3 | Cartilage, bone, umbilical cord, intervertebral disk, heart valve |
| Dermatan sulfate | 15–40 | 2.0–2.2 | Skin, heart valve, cartilage, intervertebral disk, bone |
| Keratan sulfate | 4–19 | 0.9–1.8 | Cornea, cartilage, intervertebral disk, bone |
| Heparan sulfate | 50 | 1.1–2.8 | Lung, liver, all cells? |
| Heparin | 4–16 | 3–4 | Lung, liver, skin, intestinal mucosa |

[a]From Comper and Laurent (1978).
[b]Variation in charge density essentially resides in the sulfate content.

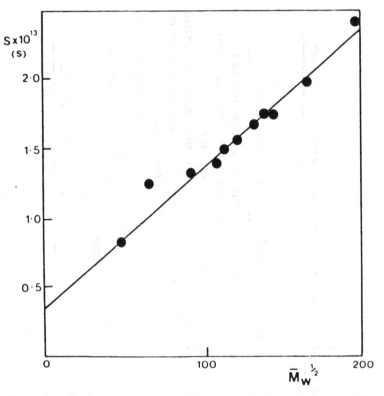

Fig. 3. Sedimentation coefficients of chondroitin sulfate
fractions as a function of the square root of molecular weight
(Wasteson, 1971).

theoretical proposals. It is a little difficult to see to what end
this labor is directed. The behavior of heparin in increasing
electrolyte concentration follows the major expectation of a
polyelectrolyte of high charge density and the clever use of
controlled desulfation confirms the main findings that the
fewer the sulfate group substitution, the less the repulsive
changes. The partial specific volume increases with
desulfation presumably because there is less electro-
striction of water, whereas the frictional coefficient, as seen
in sedimentation and viscosity experiments, falls with decreasing charge and the diffusion coefficient increases with
decreased extension of the chain. *Pari passu,* the root-
mean-square end-to-end distance and radius of gyration

fall. What is sadly not clear is the relevance of these findings to the anticoagulant or antilipemic activities of heparin.

One area in which glycosaminoglycan study has profited by hydrodynamic study is in the exploration of the self-association found in certain copolymeric, iduronate-containing polysaccharide chains. Fransson (1976) showed that this self-association was most marked among copolymers that contained similar proportions of iduronate and glucuronate and that the interaction was disturbed by chaotropic solvents such as guanidine hydrochloride. Later, Fransson et al. (1979) studied the interaction using sedimentation, viscometric, and light scattering techniques. Superaggregates were observed in these systems, disruptable either by raising salt concentration or increasing the shear rate. Superimposed on these complications was the indication of dimerization in the copolymeric materials.

The ionic environment has not been the subject of much detailed enquiry. Sheehan (personal communication) has demonstrated by viscometric techniques significant differences in the particle size of glycosaminoglycans depending on whether 0.15$M$ KCl or NaCl was the solvent.

## Proteoglycans

The transition to this parental molecule requires an effort of thought. The structure is very complex and we are only just on the verge of discovering that the spectrum of proteoglycans is a very large one indeed, probably offering a continuum of types from the epithelial mucin through to the cartilage-derived complex "bottle-brush" structure so well documented in the literature. Indeed, we may have become overobsessed with this latter macromolecule. Figure 4 shows a schematic representation of a number of established proteoglycans, where it can be seen that only the most vague generalization can be made to describe this group structurally, such as that most appear to have a bare zone, arguably substituted with oligosaccharide, separated from the glycan chain functionality. Some of these proteoglycans aggregate in the presence of hyaluronate to form complexes where many tens of monomers are individ-

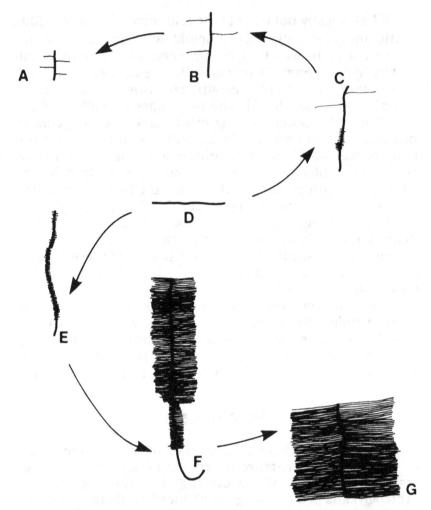

Fig. 4.   Scheme of the presently known proteoglycans, and a highly schematized suggestion of their relation to each other: (A) Heparan sulfate proteoglycan; (B) scleral dermatan sulfate proteoglycan; (C) dermal dermatan sulfate proteoglycan; (D) naked glycosylatable structural protein parent; (E) submandibular mucin; (F) cartilage proteoglycan; (G) macromolecular heparin from rat skin.

ually spaced along the hyaluronate backbone. These structures have been visualized by electron microscopy in the case of cartilage-derived material, and the original work of Sajdera and Hascall (1969b) demonstrated the sedimentation profiles obtainable for aggregates and monomers respectively.

So much chemical information is known about the cartilage proteoglycan macromolecule that our first interest will be in its description. Recently some concern has been expressed over the heterogeneity of the proteoglycan monomers in cartilage, and subtle (and sometimes tedious) methodologies have been proposed to disclose this feature. This polydispersity is intrinsic and must not be confused with problems associated with aggregatability.

Two papers form convenient twins to begin the discussion. Heinegård (1977) discusses the chemical composition of bovine nasal septa proteoglycan monomers that were subfractionated on a dissociative (i.e., containing $4M$ guanidine hydrochloride) CsCl gradient. A number of fractions of differing buoyant densities ($\rho \approx 1.60$–$1.52$) were isolated by isopycnic density gradient centrifugation, and using Sepharose 2B column chromatography and chemical derivatization the following details emerge. The polydispersity of the proteoglycan monomer was apparent in the size of the monomer, which decreased with decreasing density. The relative proportion of chondroitin sulfate, protein, and keratan sulfate-rich region increased as did the hyaluronate-binding region, whereas the (glycine and serine)-rich portion, which occurs in the chondroitin sulfate-protein linkage region, decreased. It is clear that we are dealing with a complex type of polydispersity in which no single property of the macrosystem changes unimodally.

The hydrodynamic work which complements this chemical statement is found in the paper of Kitchen and Cleland (1978). Here, bovine nasal septa proteoglycans isolated from the densest portion of a dissociative density gradient ultracentrifuge run were fractionated on Sepharose 2B, and molecular weights were determined by light scattering with single concentration experiments. The molecular weights reported range from 0.8 to 2.6 $\times$ $10^6$ and root-mean-square radii of gyration range from 35 to 52 nm. Data from this work are presented in Table 4, and show one interesting fact. The ratio of $\overline{M}_w/S_o$ remains remarkably constant and this is odd since $M/S = f/(1 - \bar{v}\rho)$.

Unless the buoyant density term is exactly compensating for the frictional coefficient, it looks as though a

TABLE 4
Solution Properties of Proteoglycan Fractions in
4*M* Guanidine Hydrochloride[a]

| $\bar{M}_w \times 10^{-6}$ | $<R_G^2>_z,$ nm$^2$ | $[\eta],$ mL g$^{-1}$ | $S_o \times 10^{13},$ s |
|---|---|---|---|
| 2.5 | 2680 | (185) | |
| (2.3) | | 179 | 26 |
| 2.2 | 2520 | (175) | |
| 2.1 | 2380 | (173) | |
| 1.8 | 1930 | (163) | |
| (1.7) | | 160 | 20 |
| 1.6 | 1840 | (157) | |
| 1.5 | 1440 | (145) | |
| (1.3) | | 139 | 15 |
| 1.1 | 1280 | 132 | 12 |
| 0.9 | 1410 | 125 | |
| 0.8 | 3300 | 120 | |

[a]All values in parenthesis are interpolated. From
Kitchen and Cleland (1978).

threefold change in size is not accompanied by any change
in the frictional coefficient. The intrinsic viscosity, however, does change roughly as expected.

The molecular weight of the protein core was found to
be $1.2 \times 10^5$ and the keratan sulfate total weight to be 1.8
$\times 10^5$, both independent of macromolecular weight, in
agreement with the findings of Heinegård. Using a treatment pioneered by Casassa and Berry (1966) for calculating
scattering factors for selected comb-branched structures
where the intersegmental distance obeys Gaussian distributions, Kitchen and Cleland (1978) show that it is possible
to "see" the chondroitin sulfate chain as shown in Fig. 5.
The agreement with Heinegård's chemical study is
heartening.

The problem of polydispersity and heterogeneity of
proteoglycans is difficult to resolve and obstructs much valuable work on the nature of the further interactions entered
into by proteoglycans with hyaluronate, link proteins,
fibronectin, and so on. Most workers are content to describe the behavior of proteoglycans in sedimentation veloc-

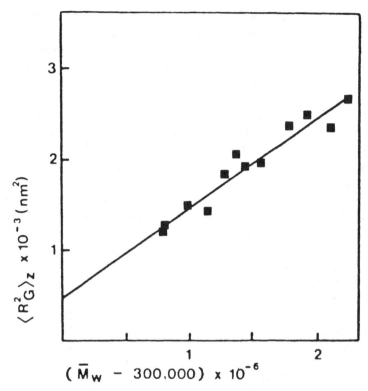

Fig. 5.  Variation in mean-square radius of gyration $< R_G^2 >$ of proteoglycan monomer fractions with molecular weight of chondroitin sulfate present.

ity experiments by sedimentation coefficients. Thus $S_o$ values of 25 S (estimated molecular weight, $2.5 \times 10^6$) are considered to represent the monomer species, whereas, in the presence of appropriate factors, $S_o$ values of 100 S (equivalent to molecular weights of $30 \times 10^6$) are ascribed to aggregates. This is not very helpful, and it is worth remembering that Sajdera and Hascall (1969b) applied an anlysis of the weight distribution function of the sedimentation data $g(S)$. Though it is difficult to extrapolate experimentally determined $g(S)$ to limiting conditions to obtain $g(S_o)$, the weight distribution function does give a measure of the weight-average molecular weight distribution in any sample, provided that suitable calibration of the relationship between $S_o$ and $M$ has been priorly found. It is an underused method of presenting data in this field.

Though this is one way in which molecular weight distribution can be quantified there is another problem that is seen if a further allowance is made for polydispersity of buoyant density. This is a real issue since Kitchen and Cleland (1978) reported a constancy for the core protein portion of proteoglycans independent of the size of the proteoglycan, and Fig. 5 reports the effect of increasing the chondroitin sulfate chain size on molecular weight. Thus there may be a natural heterogeneity of $\bar{v}$ that would reflect itself in isopycnic sedimentation techniques.

An attempt was made by Nieduszynski et al. (1980) to use fractionated samples of proteoglycans to study the interaction of those with highly calibrated hyaluronate fractions. This work shows one further problem that is met as the complexity of the system examined increases. Not all of the proteoglycans necessarily are able to bind to the hyaluronate, and it is necessary to define a term, the aggregatability (the proportion of the proteoglycan molecules capable of binding to hyaluronate) to measure this aberration. Nevertheless the paper shows how far multicomponent work can proceed with carefully fractionated material. A theoretical fit to the sedimentation data of Hardingham (1979) gave a value for the intrinsic dissociation constant for the hyaluronate–proteoglycan association of $2 \times 10^{-7}M$, in reasonable agreement with the figure of $5 \times 10^{-8}M$ found in similar salt conditions by equilibrium dialysis.

As though this was not enough complexity, Sheehan et al. (1978) report on a spontaneous self-association of pig laryngeal proteoglycan subunits that may create a further obfuscation to experiments directed to understanding the stoichiometry, reaction pathway, and mechanism of formation of the connective tissue matrix. The association, a putative dimerization, involves the protein core of the proteoglycan and has been confirmed by Hawkins et al. (1978).

## Conclusions

Much of the foregoing has sounded a pessimistic note. Study of such complex structures with their different composition of protein and polysaccharide, their high charge density, and their heterogeneous components would make this a difficult field at the best of times. Though there is little euphoria, there is some pride to be taken in seeing the extent to which a few brave souls have fought the darkness and emerged with some spoils.

The future has some pointers to it. Techniques can now be assessed for their information content and some old ones, such as sedimentation equilibrium, have much still to reveal of the distribution of molecular weights and interactional coefficients. Newer techniques, such as quasi-elastic light scattering, can yield z-average diffusion coefficients in a matter of seconds [see, for example, the work of Reihanian et al. (1979)] and can also show heterogeneities in that value (Brehm and Bloomfield, 1975). Viscosity, sadly, is the technique that, after the revolutionary appraisals of Simha in the 1950s (Simha, 1949; Weissberg et al., 1951), is in need of some refurbishment if it is to produce data capable of molecular interpretation.

## Chapter 3

# X-Ray Diffraction Analyses of Glycosaminoglycans

## Struther Arnott and Alok K. Mitra

Department of Biological Sciences, Purdue University, West Lafayette, Indiana, USA

### The Relevance of Fiber Diffraction Studies

In a uniaxially oriented fiber of a glycosaminoglycan, the molecules are extended and have symmetrical (usually helical) secondary structures. The long axes of the polymers are more or less parallel. Frequently, the fibers are also polycrystalline with clumps of molecules incorporated into small crystallites, where they make orderly lateral interactions with one another. Although this organization is artificial, its details may help illuminate the state of glycosaminoglycan molecules in solutions and tissues.

It is becoming clear that ordered secondary structures are involved in generating many physical, chemical, and biological properties peculiar to the glycosaminoglycans. These structures may not be particularly robust, nor have very great persistence lengths nor half-lives in isolation, but it would be surprising if they were not essentially the same as the structures trapped in fibers and stabilized by opportunistic lateral interactions. X-ray diffraction analysis of these fibers is at present the only means by which we can

41

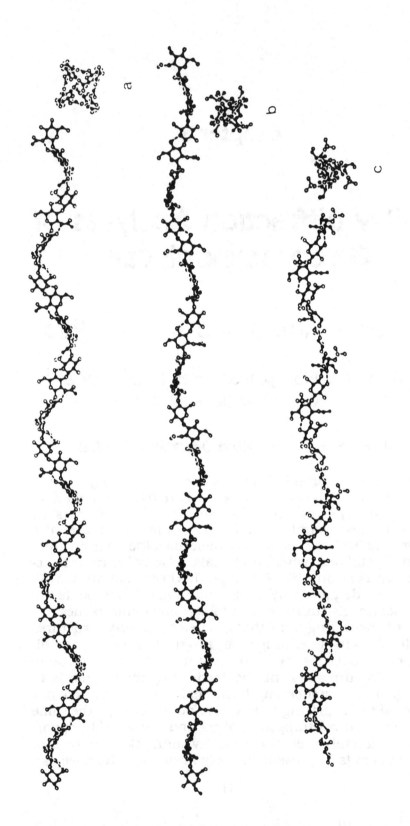

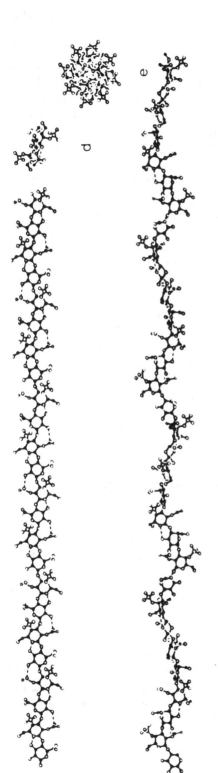

Fig. 1. Mutually perpendicular views for representatives of each of the five allomorphic helices observed with the glycosaminoglycans. The $4_3$ allomorphs with (a) $h = 0.82$–$0.85$ nm or (b) $h = 0.94$ nm are unique to hyaluronate. The $3_2$ allomorph ($h = 0.94$–$0.95$ nm), represented by (c) chondroitin 4-sulfate, is observed also with hyaluronate, chondroitin 6-sulfate, and dermatan sulfate. The $2_1$ allomorph ($h = 0.93$–$0.98$ nm), represented by (d) chondroitin 4-sulfate, is observed with every glycosaminoglycan. The $8_3$ allomorph ($h = 0.92$–$0.98$ nm), represented by (e) dermatan sulfate is observed also with chondroitin 6-sulfate.

have an accurate visualization of a highly hydrated polymer system at atomic resolution. Analysis of a system that is also polycrystalline provides not only the architecture of individual molecules, but also the details of the intermolecular interactions that may be the same as those which lead to cooperative phenomena *in aquo* and in vivo.

Thus, from the analysis of sodium chondroitin 4-sulfate (Winter et al., 1978), for example, comes first an accurate image of the threefold polyanion helices that may remedy the erroneous spatial impressions given by otherwise useful schematic formulae and identifies some of the sources of this conformation's stability as intramolecular hydrogen bonds (Fig. 1c). Second, comes a precise picture (Fig. 2) of how these polymers might interact with one another, either directly through hydrogen bonds or indirectly through water and cation bridges. Hence, one can exploit the crystal structure directly to provide a model of how the glycan chains on neighboring proteoglycan molecules might interdigitate. Or one might adapt it to visualize how parallel glycan chains on the same proteoglycan might interact at a distance.

The function of fiber diffraction studies, therefore, can be to prompt and facilitate and constrain modeling of complex systems by providing physically plausible, tangible, and precise components.

## Accuracy and Precision in Fiber Diffraction Analyses

Anyone exploiting a physical probe of molecular structure is under some obligation to provide an assessment of the accuracy of its results and some estimate of their precision. This obligation is more pressing than usual in the case of fiber diffraction studies, which are undertaken by only a few among whom there is no consensus regarding the scope and limits of such studies.

What is certainly true is that the number, quality, and resolving power of the X-ray diffraction intensities from fibrous specimens are rarely sufficient for the relative atomic positions in the diffracting molecules to be established independently with useful accuracy. However, there are systematic schemes (Fig. 3) for augmenting these data with

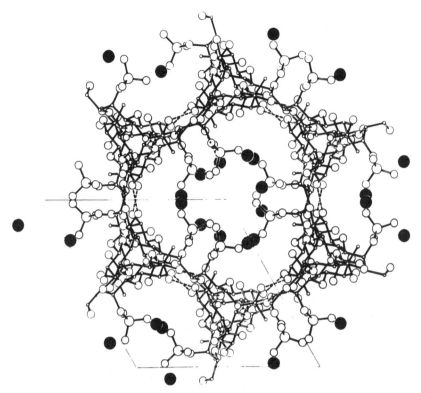

Fig. 2. A segment of the crystal structure of the sodium salt of chondroitin 4-sulfate viewed down the long axes of the polyanion chains. Around each hexagon of molecules, the chain polarity is alternately "up" and "down." The channels are filled with fully hydrated $Na^+$ ions that neutralize the carboxylate and sulfate groups protruding into the spaces.

noncontroversial stereochemical information, including certainly the primary structure of the polymer and the most probable values of its bond lengths and angles. Further metrical constraints may be provided by the dimensions and symmetry of the unit cell, by the requirements that nonbonded atoms should never be less than certain distances apart, and by the requirement that hydrogen-bonded and polar interactions should be characterized by a narrow range of distances. The melding together of these rather different kinds of data can lead to a very detailed structure in which most of the atomic positions are defined to within a few hundredths of a nanometer, which is a pre-

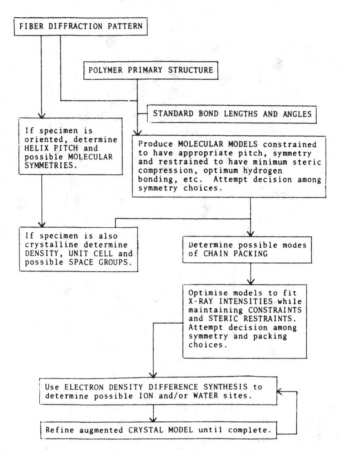

Fig. 3. A scheme for defining and refining fibrous structures using X-ray diffraction data augmented by stereochemical information.

cision adequate for identifying the critical interactions within and between molecules.

How far one can proceed varies from case to case. Since keratan sulfate specimens are uniaxially oriented but not polycrystalline, only a limited analysis is possible with curent technology. The pitch of the molecular helix (1.86 nm) is given by the layer line spacings. The helix symmetry $(2_1)$ is indicated by the meridional intensity nodes on the even layer lines (Fig. 4d). With this information one can construct a sterically reasonable model with the correct pitch and symmetry. Thereafter, one can calculate the cylindrically averaged Fourier transform of the molecule, with the hope of finding that it corresponds to the observed intensity distribution, which would indicate that the structure generated was reasonable.

From fibers of, say, calcium chondroitin 4-sulfate, which are polycrystalline as well as uniaxially oriented, one can obtain a Bragg diffraction pattern and determine a unit cell. After constructing a plausible molecular model, one can go on to determine the positions of the polyanions in this unit cell and to refine these and the molecular conformation to fit structure amplitudes derived from the Bragg intensities. Finally, one can complete the crystal structure by finding the positions of cations and water molecules. This is done with the aid of electron-density difference syntheses interleaved with rounds of re-refinement of the accumulated structure by least-squares or other optimization procedures.

For the first difference synthesis, the phases for the Fourier sum come from the polyanion crystal model, while the amplitudes are the differences between those observed and those calculated from the model. What the difference density should show is how the provisional crystal model might be improved. When this process in the case of, say, calcium chondroitin 4-sulfate, indeed yields blobs of density (Fig. 5a) of the right number to represent calcium ions and in credible positions near carboxylate or sulfate groups, there is not only an opportunity to augment the crystal model, but also the not unimportant assurance that the partial structure used in the extrapolation is essentially correct.

After least-squares refinement of the augmented structure, a new difference synthesis is calculated, removing the contributions of the polyanions and $Ca^{2+}$ ions. The remaining blobs of density (Fig. 5b), 0.25 nm or so distant from one another or from cations or oxygen or nitrogen atoms in the polyanion, are likely candidates to represent water molecules that can be incorporated into the structure, then have their positions refined and tested for steric credibility.

Eventually, a complete model for the structure emerges that has the ring of truth: in it every potential donor atom to a hydrogen bond is found to be indeed a donor; every $Ca^{2+}$ sits in a negative pocket provided by the $—OSO_3^-$

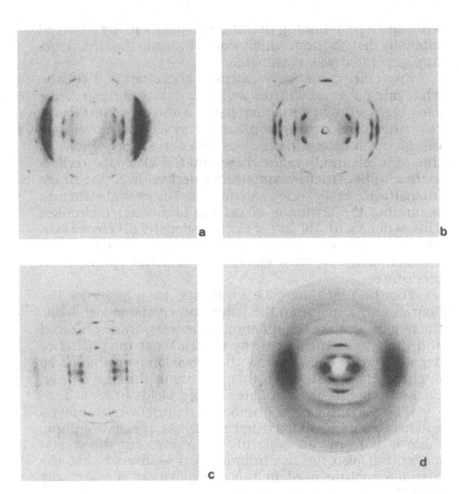

Fig. 4a–d.   The different kinds of diffraction patterns obtained from oriented glycosaminoglycan systems: (a) sodium hyaluronate with layer line spacings (molecular pitch) of 3.40 nm and intensity nodes on the meridian of the pattern every fourth layer line indicating fourfold screw symmetry (in this case $4_3$); (b) potassium hyaluronate with pitch 3.76 nm and symmetry $4_3$; (c) sodium chondroitin 4-sulfate with pitch 2.82 nm and symmetry $3_2$; (d) sodium keratan sulfate with pitch 1.86 nm and symmetry $2_1$; (e) sodium chondroitin 6-sulfate with pitch 7.84 nm and symmetry $8_3$. Patterns (a), (b), and (c) are examples of those obtained from samples that are polycrystalline as well as uniaxially oriented. Pattern (d) is an example of a merely oriented specimen.

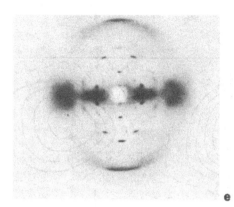

Fig. 4e.   Pattern (e) is from material of intermediate order.

and —$CO_2^-$ substituents on the chondroitin chains (Fig. 6); every $Ca^{2+}$ is coordinated to eight oxygen atoms; the number of water molecules making at least two bonding connections with other atoms totals the seven per disaccharide predicted by the density measurements.

A major flaw in this procedure is that the initial solution of the phase problem comes from a postulated model, and it is not always the case that the process of refinement can transform an erroneous model into a refineable one of the correct type. Typical dilemmas are whether the molecular helix has a left- or right-handed screw, or whether, when there are two molecules per unit cell, the molecules are parallel or antiparallel. Fortunately, there is a straightforward albeit tedious solution: analysis and refinement of all candidate models has to be conducted until the fit with the X-ray amplitudes or steric factors allows one model to be declared significantly superior to the others by some standard statistical test. Fortunately, the major dilemmas mentioned are usually resolved before the expensive process of filling in the final details of the structure has to be undertaken.

The fact that no fibrous nucleic acid structure produced by a laboratory not of Maurice Wilkins' school has survived critical re-examination is eloquent testimony to the need to invest effort in alternative structures: the model for B-DNA by Crick and Watson (1954) turned out to be a model for a member of the A family not analyzed until 20 yr

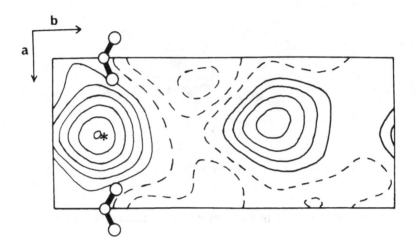

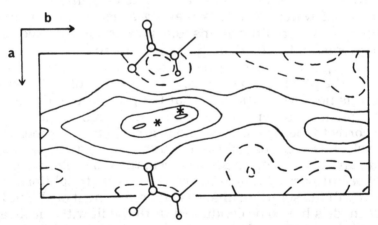

Fig. 5.  (a) A section through a putative $Ca^{2+}$ ion position in a three-dimensional map of the electron density of the hydrated calcium salt of chondroitin 4-sulfate. The contributions of the polyanions have been subtracted to enhance the other structural features. (b) Another section of the difference electron density map after removal also of the $Ca^{2+}$ ion contributions. It shows the putative positions of two neighboring water molecules.

later (Arnott et al., 1980); the three-stranded model for polyinosinic acid (Rich, 1958) should have been four-stranded (Arnott et al., 1974a; Chou et al., 1977); the double-stranded model (Langridge and Rich, 1963) for polycytidylic acid should have been single-stranded (Arnott

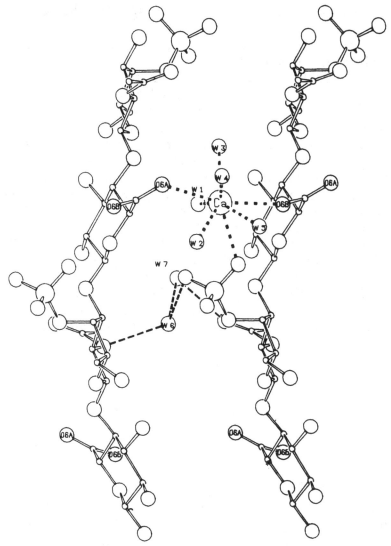

Fig. 6. A close-up of the $Ca^{2+}$ site in calcium chondroitin 4-sulfate showing that this ion bridges two carboxylate groups intermolecularly and also a carboxylate and a sulfate group of the same molecule.

et al., 1976); Mitsui et al. (1970) produced a left-handed model for D-DNA which is in fact right-handed (Arnott et al., 1974b). The point is not that one can easily be wrong in modeling a fibrous structure, but that with today's technology scrupulously applied, most gross errors are detectable.

## Polymorphism of Glycosaminoglycan Helices

By the early 1970s, all the diequatorially linked polydisaccharides under discussion had been persuaded to form uniaxially oriented and often polycrystalline fibers (Atkins and Sheehan, 1972, 1973; Dea et al., 1972, 1973; Atkins et al., 1972; Arnott et al., 1973a, 1973b, 1974c; Atkins and Isaac, 1973; Isaac and Atkins, 1973). Facile products of their diffraction patterns were the axial translation per disaccharide and possible helix symmetries. Preliminary models built with these constraints indicated that all allomorphs (with the exception of the achiral $2_1$ helices) probably had left-handed screw symmetry. This has been confirmed in detailed structure analyses of the fourfold hyaluronate helices (Guss et al., 1975; Mitra et al., 1983a, 1983b), and the threefold helices of hyaluronate (Winter et al., 1975; Winter and Arnott, 1977) and chondroitin 4-sulfate (Winter et al., 1978). The eightfold helices of dermatan sulfate, on the other hand, are almost certainly right-handed $(8_3)$ (Mitra et al., 1983c) rather than left-handed $(8_5)$.

TABLE 1
The Variety of Glycosaminoglycan Helices Trapped in
Oriented Fibers

| Glycosaminoglycan | Average axial periodicity per disaccharide, $h$ (nm)[a], for each allomorph of screw symmetry $p_q$ | | | |
|---|---|---|---|---|
| | $4_3$ | $3_2$ | $2_1$ | $8_3$ |
| Hyaluronate | 0.82–0.89[b]  0.94 | 0.94 | 0.98 | — |
| Chondroitin 6-sulfate | — — | 0.95[c] | 0.93 | 0.98[d] |
| Chondroitin 4-sulfate | — — | 0.94 | 0.98 | — |
| Dermatan sulfate | — — | 0.94[c] | 0.94 | 0.92 |
| Keratan sulfate | — — | — | 0.93 | — |

[a]Helix pitch = $h \times p$ (nm); average rotation $(t)$ per disaccharide = $360° \times (q/p)$.

[b]The pitch of the "low $h$" hyaluronate helices varies somewhat with the cation species present.

[c]The threefold helices of chondroitin 6-sulfate and of dermatan sulfate are assumed to be left-handed like those of hyaluronate and chondroitin 4-sulfate.

[d]The eightfold helix of chondroitin 6-sulfate has been assumed to be right-handed like that of dermatan sulfate.

These studies of the early 1970s defined the range of glycosaminoglycan secondary structures. This is probably complete except for the excursions available to dermatan sulfate when its iduronate residues have the energetically just as favorable $^1C_4$ chair conformation. The common covalent linkages are reflected (Table 1) in the very similar twofold allomorphs observed for all the glycosaminoglycans, in the similar threefold allomorphs observed for all except keratan sulfate, and in the eightfold allomorphs observed with chondroitin 6-sulfate and with dermatan sulfate. On the other hand, the glycosaminoglycans with the more unusual primary structures usually (but not always) are more idiosyncratic in their secondary structures. Keratan sulfate, which has no uronate residue, is observed only as the $2_1$ helix. Hyaluronate, which is quite unsulfated, has (uniquely) two types of $4_3$ allomorph, one with axial translation per disaccharide of the common extended ($h > 0.9$ nm) type, and the other with this markedly contracted ($h < 0.85$ nm). Somewhat surprisingly, the dermatan sulfate allomorphs mimic those of the chondroitin sulfates, indicating that iduronate in the fibrous systems studied has the glucuronate-like $^4C_1$ chair conformation.

Isosymmetrical allomorphs are not equally stable in the same circumstances. For example, when the countercations are exclusively sodium, the $3_2$ allomorph of chondroitin 4-sulfate is stable (Winter et al., 1978), but it is the $8_3$ allomorph of chondroitin 6-sulfate (Cael et al., 1978) and one of the $4_3$ allomorphs of hyaluronate (Winter and Arnott, 1977) which are stable under these conditions. Clearly, a better understanding of the nature of the conformational determinants could come only from detailed diffraction analyses that would reveal the organization of water molecules and cations associated with each allomorph.

## Detailed Structural Studies of Hyaluronate Allomorphs

Hyaluronate, which is not sulfated and has a relatively enormous molecular weight, holds a special place among glycosaminoglycans. Not surprisingly, therefore, it was the subject of the earliest detailed studies, with attention de-

voted mainly to its conformations with the low axial translation per disaccharide ($h \approx 0.85$ nm). These are rather different from the conformations of other glycosaminoglycans,
which all have $h = 0.9$ nm.

The screw symmetry associated with the contracted
forms of hyaluronate is $4_3$, so that the rotation per disaccharide accompanying the axial translation of 0.85 nm is
$-90°$. With these constraints, it is possible to construct
hyaluronate helices in which every glycosidic bridge has its
conformation reinforced by two hydrogen bonds (Fig. 7).
Across the (glucuronate)1 →3 (glucosamine) bridge, they
involve O2 $\cdots$ O7 (carbonyl) and (anomeric) O5 $\cdots$ O4.
Across the (glucosamine)1 →4(glucuronate) bridge, they involve N $\cdots$ O6(carboxylate) and (anomeric)O5 $\cdots$ O3.
These hydrogen bonds could reduce the chemical relactivity
of the functions involved, and help to make this
hyaluronate secondary structure relatively stiff and pos-
sessed, therefore, of anomalous hydrodynamic properties.
This is not to claim that this particular allomorph is the
only source of hyaluronate's peculiar physical and chemical
properties; but it is notable that this conformation is pre-
served essentially unchanged in a number of quite different
environments and associations where the hyaluronate molecules may be dehydrated or hydrated (Guss et al., 1975;
Mitra et al., 1983a, 1983b) or unprotonated or
hemiprotonated (Sheehan *et al.*, 1977).

Neutral sodium hyaluronate can be dry-spun to give
uniaxially oriented, polycrystalline fibers. When these are
dehydrated, the molecules nestle together in a tetragonal
crystal form where certainly the (glucosamine)O5 $\cdots$
O3(glucuronate), and possibly also the (glucuronate)O2 $\cdots$
O7(glucosamine) intramolecular hydrogen bonds are aban-
doned in favor of intermolecular associations, so that every
potential hydrogen bond donor atom is indeed a donor.
Only modest changes in individual conformation angles are
needed to convert the hypothetical helices with their maximum intramolecular hydrogen bonding to the packed helices in this dehydrated crystal form.

When the fibers are rehydrated with at least two water
molecules per tetrasaccharide, the *a* axis increases by 17%
and every *second* dissacharide assumes the form that is
doubly hydrogen-bonded intramolecularly. Although the

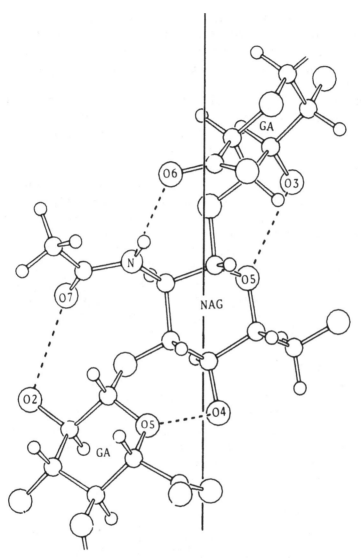

Fig. 7. The disaccharide unit of hyaluronate doubly hydrogen-bonded across both the 1 →3 and 1 →4 glycosidic linkages.

crystal symmetry is now orthorhombic rather than tetragonal, and the molecular symmetry is $2_1$ with a tetrasaccharide asymmetric unit rather than $4_3$ with a disaccharide unit, the pitch is still $4 \times 0.85 = 3.4$ nm and the molecular morphology not greatly perturbed.

An analogous situation exists when the counterion is potassium. The contracted ($h = 0.89$ nm) orthorhombic form is formally similar to that observed with sodium, although the packing of the chains and the nature of the cation pockets are surprisingly different (Mitra et al., 1983a). Upon dehydration and rehydration the tetragonal structure is obtained where the polyanion chains are extended ($h = 0.94$ nm) but, interestingly, the nature of the cation environment is similar to those seen in the orthorhombic form (Mitra et al., 1983a).

The robustness of the contracted $4_3$ conformation is further confirmed by a variant that can be trapped in fibers prepared from hyaluronic acid half-neutralized with monovalent counter cations such as $K^+$, $NH_4^+$, or $Rb^+$. This form is unique among glycosaminoglycans in having two antiparallel chains plectonemically wound in the manner of DNA. In this duplex carboxyl groups from different chains approach one another and hemiprotonation allows the formation of a carboxyl–carboxylate hydrogen bond that stabilizes the double helix (Arnott et al., 1983), as shown in Fig. 8.

There are hyaluronate allomorphs with more extended ($0.92$ nm $\leq h \leq 0.98$ nm) chains and a variety of symmetries ($4_3$, $3_2$, $2_1$). The first is observed in fibers of potassium hyaluronate (Atkins et al., 1974a) after dehydration and rehydration; the second with fibers of calcium hyaluronate (Winter and Arnott, 1977) (or of sodium hyaluronate contaminated with $Ca^{2+}$ ions) (Winter et al., 1975); the third with fibers of strongly acidified hyaluronate (Atkins and Sheehan, 1973). The roles of various ions as conformational determinants will be discussed later.

Detailed crystal-structure analyses have been performed for $3_2$ helices in calcium hyaluronate (Winter and Arnott, 1977) and a calcium-contaminated sodium salt form (Winter et al., 1975). Despite the different packing of the polyanions, their conformations vary little. Intramolecular hydrogen-bonding is reduced to one at each glycosidic bridge: O5 $\cdots$ O4, (glucuronate)1→3(glucosamine) and O5 $\cdots$ O3, (glucosamine)1→4(glucuronate). In keeping with what appears to be a general rule that all potential hydrogen bond donors are indeed donors, the other donors contrive this through intermolecular hydrogen bonds (mostly to bridging water molecules).

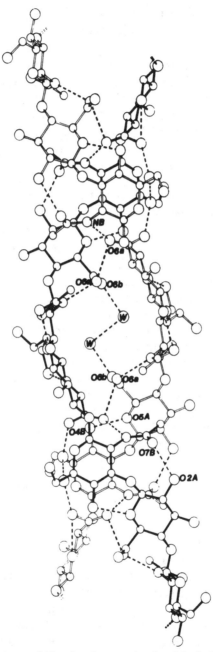

Fig. 8. A view of the hyaluronate double-helix perpendicular to the common helix axis of the antiparallel, plectonemically wound chains. The intra-duplex carboxyl–carboxylate and other hydrogen bonds are shown as broken lines. Cations have sites *between* double helices. *Within* double helices water bridges are shown: W---W.

## Detailed Structural Studies of Chondroitin Sulfate Allomorphs

Chondroitin is quite similar to hyaluronate, except that it is epimeric at the hexosamine C4. These galactosamines are usually sulfated at C4 or C6. It is perhaps this increase in (negative) charge density that disfavors the $4_3$ helices (with $t = -90°$) observed with hyaluronate and instead favors $3_2$, $2_1$, and $8_3$ helices where $t$ has numerically larger values $(-120°, -180°, -225°$, respectively). The $8_3$ (rather than $8_5$) symmetry is inferred from the detailed study of the analogous $8_3$ dermatan sulfate (Mitra et al., 1983c) since detailed quantitative analyses of chondroitin sulfates have been performed only for $2_1$ chondroitin 4-sulfate (calcium salt; Cael et al., 1978) and $3_2$ chondroitin 4-sulfate (sodium salt; Winter et al., 1978). These allomorphs have very similar conformations to their hyaluronate analogs, except that in the $3_2$ chondroitin 4-sulfate allomorph, there are only O5 $\cdots$ O3 hydrogen bonds, since the analog of the O5 $\cdots$ O4 (glucuronate)1→3(glucosamine) bond in hyaluronate is precluded by sulfation and by the configurational change.

In the crystal structures, the $3_2$ chondroitin 4-sulfate molecules form an elaborate honeycomb structure filled with fully hydrated $Na^+$ ions (Fig. 2). The $2_1$ chondroitin 4-sulfate molecules, on the other hand, form sheets of antiparallel chains. The $Ca^{2+}$ ions and water molecules are located mainly between these sheets (Fig. 9).

## Detailed Structural Studies of Dermatan Sulfate Allomorphs

Three allomorphs (with two-, three-, and eightfold) screw symmetry have been trapped in oriented, polycrystalline fibers prepared from nominally sodium dermatan sulfate. The three forms appear to be essentially isomorphous with their chondroitin sulfate and hyaluronate counterparts. It seems unlikely, therefore, that the iduronate rings in these structures can be different from the $^4C_1$ glucuronate rings in chondroitin and hyaluronate. This is not to claim that the same iduronate rings, freed from the constraints of sharing a crystal structure with unepimerized glucuronate residues would not change their shapes to $^1C_4$.

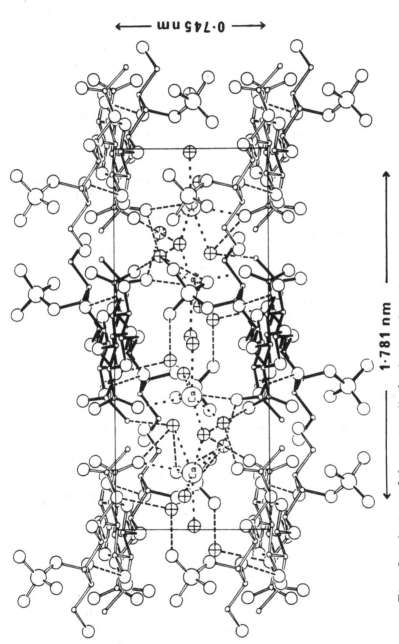

Fig. 9. A view of the unit cell of calcium chondroitin 4-sulfate looking down the long axes of the ribbon-like $2_1$ polyanions. These are packing in sheets with the neighboring polyanions antiparallel. The Ca²⁺ ions and water molecules are in the spaces between the sheets.

The results of the detailed structural studies of the eightfold helical allomorph of dermatan sulfate (Mitra et al., 1983c) reinforce the need for enterprises that permit not only the roles of water molecules and cations in the structure to be visualized, but also the accuracy of the structures proposed for the polyanions to be tested. In early modeling of isolated glycosaminoglycan chains, the allomorphs with left-handed ($4_3$, $3_2$, $8_5$) screw symmetries were invariably judged to be superior to those with right-handed ($4_1$, $3_1$, $8_3$). With the former it is always possible to make models of isolated molecules that are extensively hydrogen bonded intramolecularly. With the latter this is more difficult. Molecules in a matrix are relatively indifferent to such artificial concepts such as intra- and intermolecular hydrogen bonding. So it turns out that all the fourfold and threefold glycosaminoglycan helices in crystals are indeed left-handed (i.e., $4_3$ and $3_2$), but in the tetragonal crystal structure of dermatan sulfate, $8_3$ helices provide a better fit with the X-ray diffraction than $8_5$. They also contrive that every potential hydrogen bond donor is indeed a donor, although most of these bonds are to water molecule bridges between chains.

## The Roles of Cations in Conformation and Packing

In the first visualization of cations associated with a polyanionic polysaccharide (Guss et al., 1975), it was particularly striking how well the $Na^+$ ions nestled into pockets on each $4_3$ hyaluronate chain and could still become integrated into a larger coordination site in the crystal matrix (Fig. 10). $K^+$ ions are no exceptions, and a somewhat similar interlocking of the hyaluronate chains through an extensive scheme of cation coordination exists, e.g., in the extended $4_3$ allomorphs as shown in Fig. 11. It appears that monovalent cations play a relatively passive role in the support of glycosaminoglycan conformations and matrices. This is seen in its most extreme form in the trigonal crystal form of chondroitin 4-sulfate (Fig. 2), where the channels in the honeycomb-like matrix of polyanions are filled with fully hydrated $Na^+$ ions.

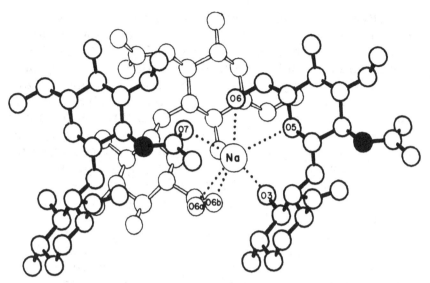

Fig. 10. A close up of the Na$^+$ ion site in the tetragonal form of hyaluronate with relatively shorter ($h$ = 0.85 nm) $4_3$ chains. The Na$^+$ ion is coordinated to three oxygens on one chain, two on another, and one on a third.

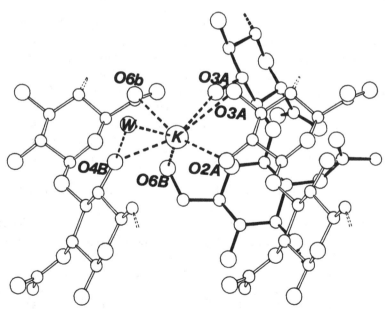

Fig. 11. A close-up of the K$^+$ ion environment in the tetragonal form of hyaluronate with extended ($h$ = 0.95 nm) $4_3$ chains. The cation coordinates to two oxygens each from the three chains and also to a water molecule.

Divalent ions (specifically $Ca^{2+}$) appear to be more active conformational determinants in the fibrous systems studied and, therefore, could have analogous effects biologically. Thus in sodium hyaluronate fibers containing only 10% $Ca^{2+}$, the hyaluronate conformation cannot retain the common $4_3$ symmetry and assumes the extended $3_2$ conformation. In the structure of calcium hyaluronate (Winter and Arnott, 1977) that contains extended $3_2$ helices the only polyanion oxygen atoms in the coordination shells of the $Ca^{2+}$ ions are carboxylate oxygens from neighboring chains. Each shell is completed by six water molecules (Fig. 12). This is the same kind of $Ca^{2+}$ environment that is observed in chondroitin 4-sulfate (Fig. 6).

It is important to recognize, however, that an effect that may be achieved by adding 10% $Ca^{2+}$ to a fiber of sodium hyaluronate, say, may require a considerably greater $Ca^{2+}$ concentration in a biological matrix where there is considerable competition for $Ca^{2+}$. Thus, X-ray studies of the calcium salts of glycosaminoglycans tell us what *can* happen in tissues in the presence of a sufficient concentration of free $Ca^{2+}$ ions, and not necessarily what *must* happen in tissues containing some $Ca^{2+}$ ions.

With chondroitin 4-sulfate the $3_2$ allomorph is the one stable in the presence of $Na^+$ ions (Winter et al., 1978), and there does not seem to be any direct coordination of charged oxygen atoms to cations. By transformation to the $2_1$ allomorph, condroitin 4-sulfate can contrive —COO⁻ $\cdots$ $Ca^{2+}$ $\cdots$ ⁻OOC—bridges between chains, analogous to those in calcium hyaluronate. This may be less important than the fact that $2_1$ chondroitin 4-sulfate can also have an intramolecular bridge —$SO_3^-$ $\cdots$ $Ca^{2+}$ $\cdots$ ⁻OOC— (Fig. 6).

Chondroitin 6-sulfate obviously cannot respond in the same way as chondroitin 4-sulfate: its $8_3$ conformation is the one observed with the $Na^+$ salt and the $3_2$ conformation with $Ca^{2+}$ (Cael et al., 1978). Since the detailed structures of these salts have not been determined, there is no precise visualization of why this is so, but inspection of molecular models suggests that $3_2$ chondroitin 6-sulfate is the only allomorph with which a —COO⁻ $\cdots$ $Ca^{2+}$ $\cdots$ ⁻$O_3S$— bridge can be made intramolecularly.

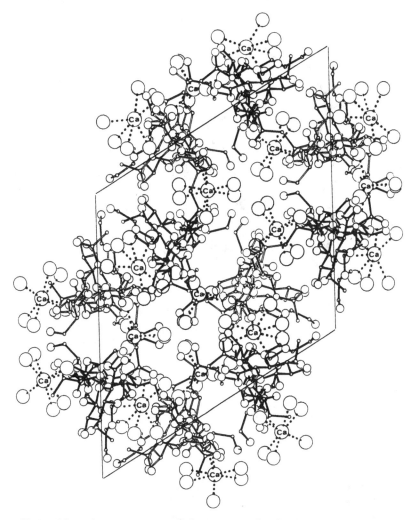

Fig. 12.   A projection of the unit cell of calcium hyaluronate. The $Ca^{2+}$ ions lie on the diad axes between antiparallel $3_2$ hyaluronate helices and bridge carboxylate groups of neighboring chains.

Dermatan sulfate conformations have not been studied systematically with cation effects in mind. The $2_1$, $3_2$, $8_3$ allomorphs are all obtained with nominally $Na^+$ salts, but it is possible that some of the specimens were contaminated with $Ca^{2+}$.

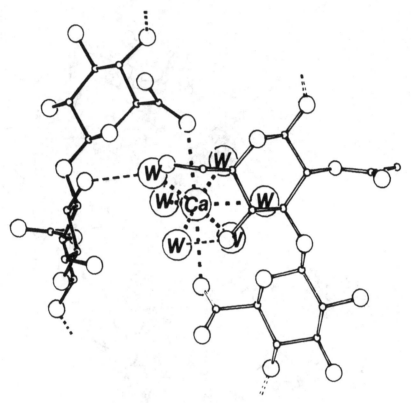

Fig. 13. Closeup of Ca$^{2+}$ coordination in calcium hyaluronate.

Strong acidification usually induces the $2_1$ allomorph in all glycosaminoglycans (Atkins and Sheehan, 1973; Atkins et al., 1972; Atkins and Isaac, 1973), but there have been no detailed visualizations that might suggest why this should be the case.

Clearly, the glycosaminoglycans respond sensitively to the cations they encounter (Table 2). The effects are often expressed finally as major conformational changes in the polyanions. The different effects of Na$^+$ and Ca$^{2+}$ are not surprising, but it is satisfying that the means by which Ca$^{2+}$ ions achieve their effects have been delineated so precisely in some instances (Winter and Arnott, 1977; Cael et al., 1978). It is also of no little interest that there are also molecular stratagems, whereby there is discrimination between Na$^+$ and K$^+$ (Mitra et al., 1983a, 1983b).

TABLE 2
Cations as Determinants of Glycosaminoglycan
Secondary Structures

| Glycosaminoglycan | Helical symmetry $(p_q)$ observed with cations shown | | |
| --- | --- | --- | --- |
| | $Na^+$ | $Ca^{2+}$ | $H^+$ |
| Hyaluronate | $4_3$ | $3_2$ | $2_1$ |
| Chondroitin 6-sulfate | $8_3$ | $3_2$ | $2_1$ |
| Chondroitin 4-sulfate | $3_2$ | $2_1$ | $2_1$ |
| Dermatan sulfate[a] | | $3_2, 2_1, 8_3$ | |
| Keratan sulfate | $2_1$ | $(2_1)^b$ | $(2_1)^b$ |

[a]The correlations have yet to be made in the case of dermatan sulfate.

[b]Keratan sulfate is assumed to be locked in a single allomorphic form.

## Proteoglycans and Heterotypic Interactions

Except for hyaluronate, the glycosaminoglycans in vivo are tethered covalently to polypeptides. Moreover, the glycosaminoglycan chains on a particular polypeptide need not all be of the same chemical type. On the other hand, the X-ray experiments have been conducted for the most part on glycosaminoglycans that have been separated from protein and fractionated to produce specimens with relatively homogeneous primary structures. Therefore, one must ask whether glycosaminoglycans covalently linked to protein in a proteoglycan behave differently from isolated glycosaminoglycan chains. One must ask also what might be the effect of having chemically different glycosaminoglycans in the same matrix.

Experiments with chondroitin 4-sulfate and its proteoglycan (Atkins et al., 1974b; Winter et al., 1978; Cael et al., 1978) show that the protein in the latter case does not prevent the polysaccharide chains from organizing themselves into precisely the same allomorphic forms that are observed in the absence of protein. Nor does the protein interfere with the effects of different cations. Moreover, some of the experiments with dermatan sulfates were conducted with "doublets" (i.e., partially digested dermatan sulfate proteoglycan), and these too gave the same results (Arnott et al., 1973a) as protein-free dermatan sulfates (Atkins and Isaac, 1973).

Thus it seems that the protein core of a proteoglycan is neither organized nor readily organizable into a regular structure. It appears that its main function may be no more than to maintain a locally high concentration of polysaccharide chains.

The biological functions or physical properties of glycosaminoglycan matrices containing more than one glycan component remain to be clarified. The proteoglycans from pig mucosa have mainly chondroitin 4-sulfate side chains, but they also contain keratan sulfate. That $2_1$ glycosaminoglycan chains were observed with a $Na^+$ salt (Atkins et al., 1974b) might be a result of some $Ca^{2+}$ ions being present. Alternatively, it might be the result of the conformationally less versatile keratan sulfate chains imposing their limited secondary structure on the whole matrix. That heterotypic interactions between glycosaminoglycans are worth considering is emphasized by the fact that hyaluronate can be induced to assume the unusual $4_3$ helix with the long axial translation per residue ($h = 0.98$ nm) in the presence of 10% chondroitin 6-sulfate. However, the properties of systems with two (or more) carbohydrate components have scarcely been explored despite these provocative preliminary observations.

## Conclusions

In the last decade X-ray diffraction analyses of oriented fibers of glycosaminoglycans have: (i) delineated the boundaries of the possible variations of their secondary structures; (ii) provided visualizations of the homotypic interactions between glycosaminoglycan chains for a number of hyaluronate and chondroitin and dermatan sulfate structures; (iii) shown in great detail the modes of association of water molecules and cations with glycosaminoglycans; (iv) uncovered the role of cations as conformational determinants and defined in a number of cases precisely how the determination is contrived.

Important areas that remain to be explored further include: (i) the iduronate-containing glucosaminoglycans, in which there is the possibility of quite different secondary structures being trapped where the iduronate rings are $^1C_4$ rather than $^4C_1$ chairs; (ii) heterotypic interactions between unlike glycosaminoglycans in two-component systems and the possible modulations of responses to cations in such systems compared to the more homogeneous ones already studied.

## Chapter 4

# Conformation of Individual Residues and Chain Segments of Glycosaminoglycans in Solution by Spectroscopic Methods

## B. Casu

Istituto di Chimica e Biochimica "G. Ronzoni,"
Milan, Italy

### Introduction

A variety of spectroscopic methods are being used for characterizing the conformation and interactions of glycosaminoglycans in solution. These methods are based on the electronic (UV-VIS), vibrational (IR-Raman) and nuclear magnetic resonance (NMR) spectra.

Although the direct use of *electronic spectra* gives very poor information, mainly because the only UV-absorbing groups present ($COO^-$ and CO—NH) are weak chromophores, chiroptical methods (ORD and CD) based on perturbation of these chromophores can provide valuable information on the conformation of the glycosaminoglycan alone or in the presence of complexing species having stronger chromophores (Stone, 1965; Morris and Sanderson, 1973; Balazs et al., 1977).

The use of *vibrational spectra*, also of great potential for conformational studies of glycosaminoglycans, is at present limited by the complexity of vibrational patterns of these polymers and is based essentially on empirical correlations. "Group frequencies" are mainly used to monitor conformational transitions or interactions with other species. Probably the only established "conformational" use of IR spectra for glycosaminoglycans is the assignment of axial or equatorial orientation of sulfate groups from frequencies in the 800 cm$^{-1}$ region. However, these correlations do not seem to be generalizable, and have not been applicable to aqueous systems, mainly because the above region is hardly accessible using $H_2O$ or $D_2O$ as solvent (a limitation, however, that is being removed by improvements in instrumentation). Raman spectroscopy is quite attractive because it can be applied using either $H_2O$ or $D_2O$ as solvent over the whole range, and provides good "fingerprints" for individual glycosaminoglycans. However, the assignment of most Raman peaks, arising from strongly coupled vibrations, is still too formidable a task (Cabassi et al., 1978; Casu et al., 1978; Perlin and Casu, 1982).

*NMR spectroscopy* is thought to be the most informative method for the conformational analysis of glycosaminoglycans, especially for assessing the conformation of individual residues or the local conformation of functional groups. As a multiparametric technique, NMR provides a variety of "handles" for investigation of detailed structure. First of all, at least two kinds of nuclei ($^1H$ and $^{13}C$) can now be currently used and others, such as $^{15}N$ and $^{17}O$, are expected to be accessible in the near future. Each nucleus type can potentially provide one signal for every non-equivalent atom, and at least three parameters (chemical shift, coupling constant, and relaxation times) per signal, the only limitation being instrumental resolution, which is also expected to improve (Perlin, 1976b; Perlin and Casu, 1982). Some applications of NMR spectra to conformational analysis of glycosaminoglycans have been reviewed (Gatti, 1978).

The maximum number of NMR parameters can at present be measured only in those favorable cases in which resolution of all the signals can be achieved. As with other biopolymers, signal overlap is usually severe for glycosaminoglycans. This results from both a close similarity of the magnetic environment of sets of atoms, and viscosity effects that cause signal broadening. Structural heterogeneity, a common feature of all glycosaminoglycans (with the probable exception of hyaluronate), not only introduces additional signals that may overlap on major ones, but also produces "sequence-effects," i.e., small shifts of signals from atoms of neighboring residues. These residues "feel" a different magnetic environment when a neighboring residue is substituted. The overall effect is a rather complex pattern. For the above reasons, an effective NMR approach to conformational analysis of glycosaminoglycans requires use of high fields (especially for the $^1$H spectra) and special resolution-enhancement techniques.

Because of the great deal of information that is being made available by NMR studies on the subject, this presentation will preferentially concentrate on the NMR approach. Results from other approaches will nevertheless be discussed and compared with those from NMR.

## Conformation of Individual Residues

Figure 1 shows a typical $^1$H spectrum of a glycosaminoglycan at 270 MHz in $D_2O$ solution. The sample was a fraction of urinary heparan sulfate—an especially homogeneous preparation of low sulfate content, with D-glucuronic acid as the major uronic acid residue (Tira et al., 1979). For simplicity, the spectrum can be divided into three regions. The first region (2–2.2 ppm) is typical for the methyl signal of N-acetyl groups. The central region (3–4.5 ppm) contains the signals from the non-anomeric ring hydrogens, and the third region (4.5–5.6 ppm) is typical for the anomeric hydrogens. The anomeric signals illustrate current criteria for assigning local conformations:

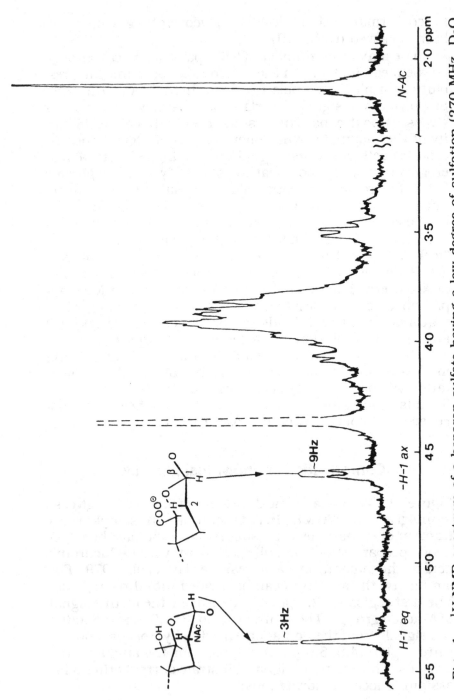

Fig. 1. $^1$H-NMR spectrum of a heparan sulfate having a low degree of sulfation (270 MHz, D$_2$O, 70°C) (Tira et al., 1979).

1. Axial C—H signals occur at higher fields than equatorial C—H (having an otherwise similar environment).

2. The separation between the components of each signal is small (~3 Hz) when the dihedral angle between the C—H bond and its vicinal C—H is ~60° (i.e., for a *gauche* arrangement) and large (~9 Hz) when the angle is ~180° (*trans–diaxial* hydrogens).

Intermediate values correspond to intermediate angles, according to the well-known Karplus equation (Perlin, 1976b; Perlin and Casu, 1982).

Although partial analysis of the ${}^1H$ spectrum, as for heparan sulfate in Fig. 1 and a recently published 400 MHz spectrum (Huckerby and Nieduszynski, 1982), is frequently used for deducing the conformation of the pyranose rings, this is reasonably reliable only when all the interproton coupling constants can be measured, i.e., when all the signals can be satisfactorily resolved.

The spectrum of N-acetylchondrosine at 270 MHz in $D_2O$ (Fig. 2) illustrates a more complete analysis. With the nonviscous solution of a small-molecular weight species, problems associated with signal broadening are greatly reduced, but the spectrum is nonetheless complex because the sample is a mixture of α- and β-anomers (produced by mutarotation of the reducing galactosamine residues). The spectrum contains a total of 24 C—H signals; most of these are well resolved, but others severely overlap even at this high magnetic field. Signals were assigned by homonuclear spin-decoupling. The spectrum is being computer-simulated for a refined calculation of the coupling constant ($J$) values (Gatti, unpublished). However, the approximate (first-order) values of these coupling constants can be obtained from the measured splittings of the signals.

There are a number of large (~9 Hz) as well as small (~3 Hz) splittings, the former attributable to *trans–diaxial* hydrogens (all those of the glucuronate residue and some of the N-acetylgalactosamine residue (1,2 of the β-anomer and 2,3 of both anomers). The spectrum is thus consistent with the conformation ${}^4C_1$ for both residues. It also nicely illustrates the "sequence effect." The anomeric hydrogens of the uronate residue absorb at slightly different field for the α- and the β-forms, i.e., they "see" the different configu-

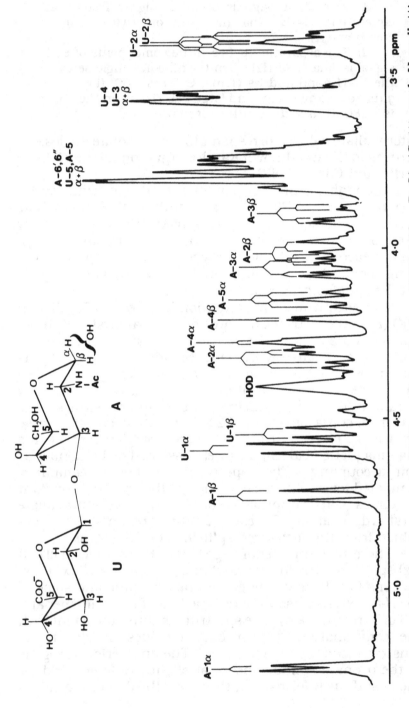

Fig. 2. ¹H-NMR spectrum of *N*-acetylchondrosine (270 MHz, D₂O) (Gatti and Vercellotti, unpublished).

ration at C(1) of the contiguous residue. Similar differences are observable for H(2) of the uronate residue. (Sequence effects are usually much more pronounced in the $^{13}$C spectra, partly because of the much wider chemical shift range than for $^{1}$H.)

Calculated values for interproton *coupling constants* are now available for hyaluronate (Welti et al., 1979), chondroitin sulfates (Gatti, 1978; Welti et al., 1979), dermatan sulfate (Gatti et al., 1979a), and heparin (Gatti et al., 1978 and 1979b).

Table 1 reports these values for hyaluronate. They are all large (7.7–10.5 Hz), in the range of *trans–diaxial* couplings and thus consistent with the $^{4}C_{1}$ conformation of both the uronic acid and the aminosugar residues. Similar conclusions were obtained for chondroitin sulfates (Gatti et al., 1979a,b; Welti et al., 1979). The coupling between H(5) and H(6') is smaller than between H(5) and H(6''), indicating a nonsymmetrical orientation of 6' and 6'' hydrogens relative to H(5).

Although the ring conformation deduced by NMR for the glucuronate and *N*-acetylhexosamine residues of hyaluronate, chondroitin sulfates, and dermatan sulfate were largely as expected because of their intrinsic stability, the preferred conformations of iduronate residues of heparin, heparan sulfate, and dermatan sulfate are not easily predictable, since the energies calculated for the two alternative forms are quite similar (Angyal, 1969). The candidate conformations for both uronate residues are shown

TABLE 1
Interproton Coupling Constants (Hz) for Sodium
Hyaluronate in $D_2O$[a]

| J | Glucuronate residue | Aminosugar residue |
|------|---------------------|--------------------|
| 1,2 | 7.7 | 8.5 |
| 2,3 | 8.5 | 9.3 |
| 3,4 | 8.5 | 9.0 |
| 4,5 | 9.0 | 10.5 |
| 5,6' | | 2.0 |
| 5,6'' | | 5.0 |

[a]From Welti et al. (1979).

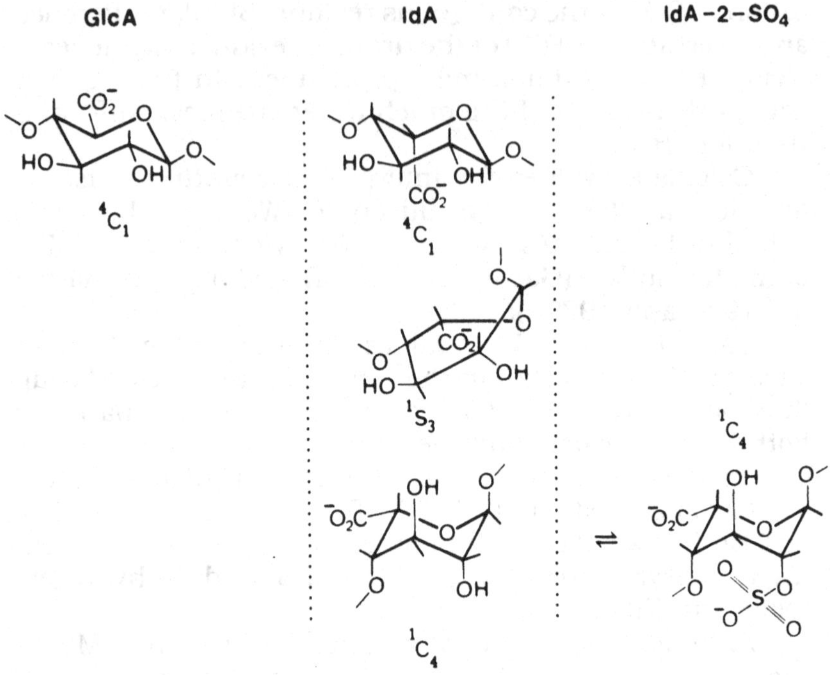

GlcA                          IdA                     IdA–2–SO$_4$

Fig. 3.   Proposed conformations for glucuronate (GlcA) and iduronate (IdA) residues (Fransson et al., 1978).

in Fig. 3 (Fransson et al., 1978). Though the conformation of D-glucuronate residues, as previously seen from the NMR data for heparan sulfate, N-acetylchondrosine, and hyaluronate, is undoubtedly $^4C_1$, it is still debated whether the conformation of L-iduronate is $^4C_1$ or its alternative form $^1C_4$. A skew-boat form ($^1S_3$) has also been proposed (Nieduszynski et al., 1977).

At first glance, NMR appears to be the method of choice for discriminating between the candidate conformations: $^4C_1$ has all *trans–diaxial* hydrogens (except 4,5), $^1C_4$ has all *gauche* hydrogens, and $^1S_3$ has two "*quasi-gauche*" and two "*quasi-trans–diaxial*" arrangements. The corresponding interproton couplings for L-iduronate should therefore be all large (except 4,5) if the conformation is $^4C_1$, all small if it is $^1C_4$ and two small (1,2 and 4,5) plus two large (2,3 and 3,4) if the conformation is $^1S_3$.

Although the $^1H$ spectrum of heparin is not completely resolved even at 270 MHz (Fig. 4a), application of

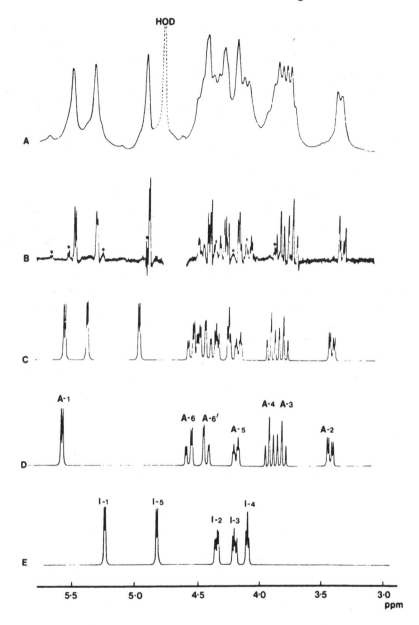

Fig. 4. ¹H-NMR spectra of beef-lung heparin (270 MHZ, D₂O, 35°C). A: normal; B: convolution-difference spectrum; C: computer-simulated; D: computer-simulated spectrum of aminosugar residues; E: computer-simulated spectrum of iduronate residues (Gatti et al., 1978).

resolution-enhancement methods such as "convolution–difference" (Campbell et al., 1973) allowed the separation of all the signals (b), which were unequivocally assigned by homonuclear spin-decoupling (Gatti et al., 1978 and 1979b). The spectra of the component monomer residues were computer-simulated (d and e) and their addition (c) closely matched the experimental spectrum. (In this heparin preparation, from beef lung, more than 85% of the uronic acid was present as L-iduronate. Signals labeled with asterisks do not belong to the regular repeating units.)

The interproton coupling constants calculated from the computer-simulated spectrum are reported in Table 2 (Gatti et al., 1978). The values for the iduronate residues are all small ( < 3.5 Hz, with the somewhat higher value of 5.9 Hz for $J_{2,3}$), indicating all *gauche* or *quasi-gauche* hydrogens. The values for the acetamidohexose residues are large (9–10 Hz) for 2,3, 3,4, and 4,5 hydrogens (indicating a *trans-diaxial* arrangement) and small only for 1,2 (*gauche* arrangement). These couplings therefore point to a $^1C_4$-like conformation of L-iduronate residues and a $^4C_1$ conformation of acetamido D-hexose residues, as previously suggested by partial analysis of the spectra at 220 MHz (Perlin et al., 1970). (The small $J_{1,2}$ value of the latter residues arises because, in an α-linked aminosugar H(1) and H(2) are *gauche*, as already seen for heparan sulfate.) $J_{5,6'}$ and $J_{5,6''}$ are both small and rather similar.

Figure 5 shows the Newman projections corresponding to the approximate dihedral angles indicated by the

TABLE 2
Interproton Coupling Constants (Hz) for Heparin in
$D_2O$ at 35°C[a]

| $J$ | Iduronate residue | Aminosugar residue |
|-----|-------------------|--------------------|
| 1,2 | 2.64 | 3.66 |
| 2,3 | 5.90 | 9.98 |
| 3,4 | 3.44 | 9.09 |
| 4,5 | 3.09 | 9.23 |
| 5,6' |  | 2.92 |
| 5,6'' |  | 2.15 |

[a]From Gatti et al. (1978).

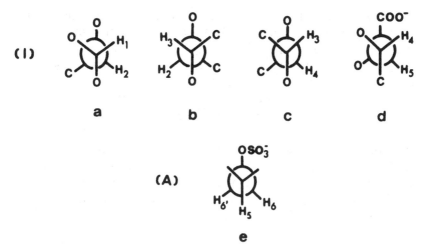

Fig. 5. Newman projections for ring protons of iduronate residues (I: a,b,c,d) and protons at C(6) of the aminosugar residues (A) of heparin (Gatti et al., 1978).

interproton couplings for the ring hydrogens of iduronate residues and for the C(6)-OH fragment of acetamidohexose residues (Gatti et al., 1978 and 1979b). With respect to iduronate residues, there are four sets of *gauche* hydrogens, one (a) having one oxygen and one carbon opposite (antiperiplanar) to the hydrogens, two (b and c) with two antiperiplanar carbons, and one (d) with two antiperiplanar oxygens. Since electronegative groups antiperiplanar to the hydrogens decrease the extent of interproton coupling, the largest values for *gauche* couplings observed in a number of simple sugars (i.e., 5.5 Hz) (De Bruyn and Anteunis, 1976) are associated with orientations such as in (b) and (c). The rather large $J_{2,3}$ coupling for iduronate residues of heparin can therefore be largely accounted for by the fact that the hydrogens are opposed by carbons. As discussed later, this does not exclude the possibility that a slight distortion of the $^1C_4$ chair also contributes to raising the coupling above the average value.

With respect to the local conformation of the C(6)-OH fragment of the acetamidohexose residues, the observed couplings indicate that the rotamer *e*, with angles of 60° between H(5) and the two C(6)—H bonds, is the most favored, probably because of the requirement to place the bulky and charged $SO_3^-$ group as far as possible outside the ring. The above orientation is in fact different from those found for glucose (Koch and Perlin, 1970), N-acetylglucosamine (Perkins et al., 1977) and N-acetylgalactosamine (Welti et al., 1979), where the C(6)—OH substituent appears to assume an asymmetrical arrangement.

Better spectra of heparin were obtained at higher temperatures, essentially because of reduced solution viscosity and increased segmental motions. Figure 6 shows the measured and computer-simulated $^1$H spectrum of heparin at 90°C. The coupling constants calculated from these spectra were essentially the same as obtained from the spectrum at 35°C, indicating no substantial change of conformation of either residue (Gatti et al., 1979b).

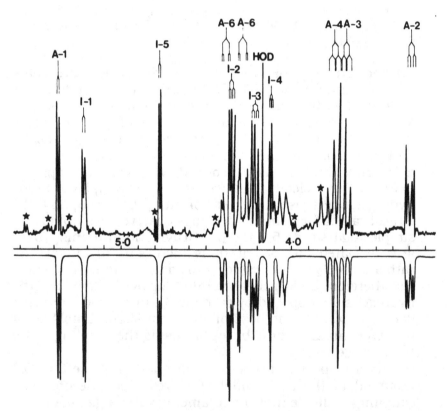

Fig. 6.   Convolution-difference $^1$H-NMR spectrum of beef-lung heparin (270 MHz, $D_2O$, 90°C) (upper trace), and the corresponding computer-simulated spectrum (lower trace). (Gatti et al., 1979b).

Information on the conformation of heparin residues was also obtained from the $^{13}C$ spectrum, especially from the magnitude of couplings between the anomeric carbons and their appended protons. Figure 7 is the $^{13}C$ spectrum of a beef-lung heparin, proton-decoupled (a) and coupled (b). The signals were assigned by heteronuclear spin decoupling (Gatti et al., 1979b). The $J_{C(1)H(1)}$ values for the anomeric carbons of both residues are of the order of 170 Hz, characteristic of axially orientated C(1)—O bonds, as expected from the $^1C_4$ conformation of α-L-iduronate and $^4C_1$ conformation of acetamido α-D-glucose (Hamer and Perlin, 1976; Fransson et al., 1978; Gatti, 1978).

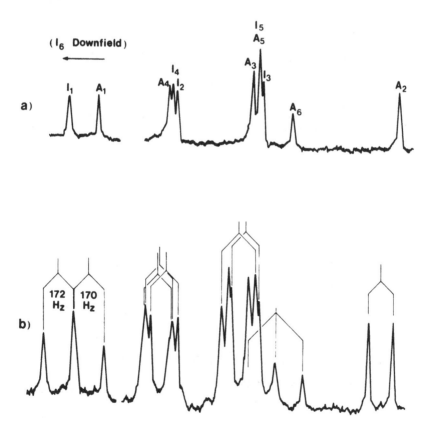

Fig. 7.    $^{13}C$-NMR spectrum of beef-lung heparin (67 MHz, $D_2O$, 35°C): (a) proton-decoupled; (b) coupled (Gatti et al., 1979b).

With respect to the nonsulfated iduronate residues of dermatan sulfate, only rather recently was it possible to obtain well-resolved $^1$H spectra suitable for measuring the interproton coupling constants. Figure 8 compares the 270 MHz convolution-difference spectrum of dermatan sulfate (b) with that of chondroitin sulfate (a). Signals were assigned as usual by homonuclear spin-decoupling. Although the couplings for the glucuronate residues of chondoitin-4

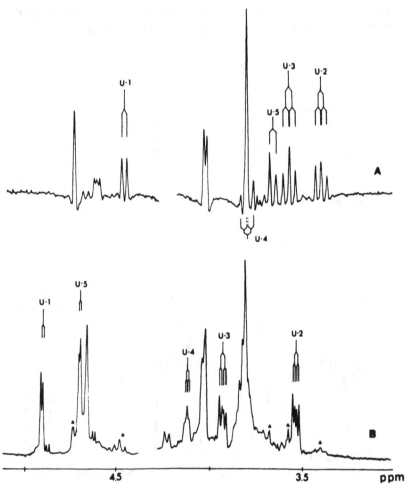

Fig. 8. Convolution-difference $^1$H-NMR spectra of chondroitin 4-sulfate (A) and dermatan sulfate (B) (Gatti et al., 1979a).

TABLE 3
Chemical Shift Data for Uronic Acid Protons of Glycosaminoglycans in $D_2O^a$

| Gycosamino-glycan | H(1) | $(J_{1,2})$ | H(2) | $(J_{2,3})$ | H(3) | $(J_{3,4})$ | H(4) | $(J_{4,5})$ | H(5) |
|---|---|---|---|---|---|---|---|---|---|
| Chondroitin 4-sulfate | 4.46 | (8.0) | 3.40 | (8.5) | 3.58 | (9.0) | 3.81 | (9.0) | 3.66 |
| Dermatan sulfate | 4.91 | (3.0) | 3.54 | (6.0) | 3.93 | (3.5) | 4.13 | (3.3) | 4.69 |
| Heparin | 5.22 | (2.64) | 4.35 | (5.9) | 4.20 | (3.44) | 4.11 | (3.09) | 4.82 |

$^a$From Gatti et al. (1979a). Chemical shifts: $\delta$, ppm from TSP; interproton coupling constants: $J$ (Hz).

sulfate are all large (see also Welti et al., 1979), L-iduronate in dermatan sulfate shows all small couplings. In fact, as shown in Table 3, which compares chemical shifts and couplings for a series of glycosaminoglycans, the couplings for dermatan sulfate and heparin are seen to be essentially the same (Gatti et al., 1979a).

Since nonsulfated iduronate residues are components of the heparin chains that contain the binding sites for antithrombin-III (Lindahl et al., 1979), the comparison of the NMR patterns of sulfated and nonsulfated iduronate residues of heparin is of obvious interest. Removal of the 2-O-sulfate group of the iduronate residues of heparin causes a dramatic downfield shift of the C(1) signal from these residues. Together with potentiometric evidence of the presence of two types of carboxyl groups, this behavior was interpreted in terms of two different forms (conformations) of iduronate residues in desulfated heparins, the relative proportions being dependent on the method of desulfation (Fransson et al., 1978). However, only one set of couplings was observed in the high-field $^1$H-NMR spectra of N,O-desulfated heparins, the most easily measurable ones ($J_{1,2}$ and $J_{4,5}$) being essentially the same as in unmodified heparin (Gatti, Casu, and Fransson, unpublished).

The couplings observed for heparin, desulfated heparin, and dermatan sulfate, indicating a $^1C_4$-like conformation for the L-iduronate residues, were also observed for the major disaccharide obtained by deaminative hydrolysis of heparin (Gatti et al., 1979b) and in the model compound methyl α-D-iduronate (Perlin et al., 1972). Also, methyl α-L-idopyranoside showed similar couplings (Gatti and Sinaÿ, unpublished). It seems, therefore, that this conformation is characteristic of idopyranose rings, irrespective of whether they carry the carboxylate or sulfate ester function, in the monomeric form or incorporated in dimeric or polymeric chains. Although some of the interproton couplings (especially $J_{2,3}$) slightly increase with increasing temperature (Perlin et al., 1972; Gatti et al., 1979b), the coupling patterns remain substantially compatible with a predominant $^1C_4$-like conformation.

These conclusions are different from those from other studies, at least as far as dermatan sulfate is concerned. In fact, the conformation indicated for L-iduronate residues of dermatan sulfate by periodate oxidation studies (Scott, 1968; Fransson, 1974; Di Ferrante et al., 1971) and X-ray diffraction studies (Atkins and Isaac, 1973; Arnott et al., 1974c) point to a $^4C_1$ conformation. Although it is conceivable that the solid-state conformation is different from that in solution, the discrepancy with periodate oxidation is not easily explainable. In fact, the fast oxidation of dermatan sulfate (Scott and Tigwell, 1975, 1978; Fransson, 1974) cannot be reconciled with the unfavorable *trans–diaxial* arrangement of the hydroxyls at C(2) and C(3) in the $^1C_4$ L-iduronate ring. Although it might be argued that the rate of oxidation is related to the conformation in the transition state, the oxidation should obviously be favored by an angle lower than 180° between the two OH groups. Such a distortion, leading to a somewhat lower angle between O(2) and O(3) is not incompatible with the observed couplings (Gatti et al., 1979a). A distorted form of this kind was convincingly demonstrated for the idopyranose ring of methyl 4,6-*O*-benzylidene-D-idopyranose in dimethylsulfoxide–$D_2O$ solution (Paulsen and Friedman, 1972).

## Conformation of Chain Segments

Since coupling between protons is not transmitted through the glycosidic bonds, coupling criteria cannot be used to assess the conformation of polysaccharide chains. A promising approach developed by Perlin and associates for di- and oligosaccharides, making use of "vicinal" $^{13}C$—$^1H$ couplings across the glycosidic bridges (Perlin, 1976a), is not yet applicable to glycosaminoglycans because of signal broadening and interference by other couplings. Disruption of these interfering couplings by selective deuteration, as is feasible for disaccharides (Hamer et al., 1978), promises to provide useful parameters in the near future.

Chemical shift criteria for determining conformations are much more risky, because of the many factors affecting this parameter. Shifts that occur as a result of changes of pH or as a function of the presence of other species in solution, without affecting the coupling pattern, can have alternative interpretations in terms of changes of polarization of ionizable groups, changes in local conformation of substituent groups on the pyranose rings, and/or changes in chain conformation. Caution must therefore be exerted in interpreting any chemical shift changes exclusively in terms of changes in chain conformation, even when the invariance of couplings provides good evidence for assuming invariant conformation of the monomeric residues. However, since the first two effects are usually (though not necessarily) confined to the immediate environment of the sites of changing polarization or local conformation of substituent groups, it can, as a first approximation, often be assumed that chemical shift changes at sites involved in the glycosidic linkages reflect changes in the interglycosidic torsional angles. Changes in these angles should in fact bring about changes in strength of secondary magnetic fields felt by nuclei as a result of magnetic anisotropy or field effects. In particular, the chemical shifts of the hydrogens at the glycosidic bridges are expected to be within the "shielding cone" of the secondary magnetic fields originated by circulating electrons of the opposing bonds, and to be progressively more deshielded as these bonds depart from the eclipsed orientation (Casu et al., 1970).

Carbon nuclei, which are less sensitive than hydrogens to magnetic anisotropy effects are, however, dramatically affected by field (so-called "steric") effects depending on orientation in space of carbons at positions $\beta$ and $\alpha$ (Perlin, 1976a).

As shown in Fig. 9 for the NMR titration curves of heparin, most $^1H$ and $^{13}C$ chemical shifts are strongly affected

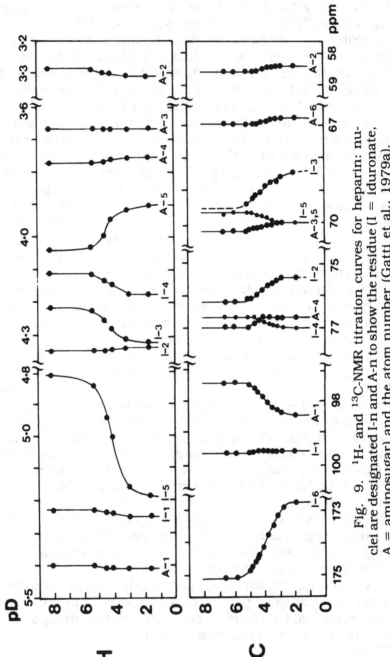

Fig. 9. $^1$H- and $^{13}$C-NMR titration curves for heparin: nuclei are designated I-n and A-n to show the residue (I = iduronate, A = aminosugar) and the atom number (Gatti et al., 1979a).

on going from neutral to acid solution. A few signals are more sensitive than others to these pH changes, since the curves for these (Fig. 9) clearly show inflexions at pH values corresponding to the $pK_a$ value of the carboxyl groups (3.9–5.1 according to concentration). It can be seen that all the proton and carbon signals of iduronate residues are affected, most notably those at the site of changing ionization [H(5) and C(6)], as expected as a result of changes in polarization, and consequent changes in diamagnetic shielding of these nuclei. This polarization is felt across several bonds [up to H(3) and C(2)].

The pH-induced shifts for hydrogens and carbons of the glycosidic bonds (at positions 1 and 4) of both iduronate and glucosamine residues are thought to be essentially the result of small changes in chain conformation (most probably a slight contraction as the ionization of the COO⁻ groups is suppressed). Most significant is the pH sensitivity of H(5) of the acetamido residue, a hydrogen quite removed from the ionization centre.

As shown in Fig. 10, the H(5) resonances from both residues experience selective broadening at pH values close to the $pK_a$, and return to "normal" shape at lower pH. As visualized in Fig. 11, H(5) of the glucosamine residue is indeed quite close in space to the carboxylate group of the adjacent iduronate residues, and is thus expected to lie within the shielding cone of this group. The selective broadening of the two H(5) signals is thought to arise from slow exchange of the carboxyl proton when, on the average, only half of the carboxyls are ionized (Gatti et al., 1979b).

The $^1H$ and $^{13}C$ resonances of heparin are shifted on addition of $Ca^{2+}$ and other ions (Perlin, 1975; Boyd et al., 1980; Liang et al., 1982), and these effects also appear to be associated with changes in chain conformation. The interproton coupling constants remain substantially unchanged in the presence of these ions (Casu et al., 1984). Striking selective broadenings of some $^1H$ and $^{13}C$ signals of heparin were observed on addition of trace amounts of the relaxation reagent, gadolinium nitrate. These broadenings, observed also for dermatan sulfate, were shown to be stereoselective in that they were observable only for uronic residues bearing axial C(1)—O bonds (Casu et al., 1975).

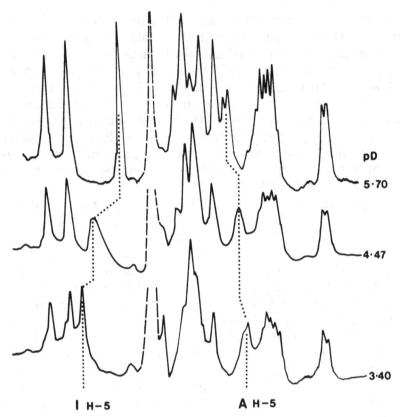

I H-5                            A H-5

Fig. 10. ¹H-NMR spectra of beef-lung heparin (270 MHz, 35°C) at different pD values (Gatti et al., 1979a).

Another transition amenable to investigation by NMR is the alkali-induced change in chain conformation of hyaluronate. Figure 12 shows the ¹H spectra (300 MHz) of low molecular weight hyaluronate in $D_2O$ and in $0.4M$ NaOD. The signals were assigned by spin-decoupling and computer-simulation (For convenience of presentation, the spectra of the figure are the calculated ones). A number of signals are dramatically and reversibly shifted at high pH, especially the H(2) signals of different residues (Welti et al., 1979). These shifts, which do not have a counterpart in model methyl glycosides, are consistent with the chain conformation for hyaluronate shown in Fig. 13, similar to that first proposed by Scott and Tigwell (1975) to account for slow rates of periodate oxidation. This "super-hydrogen-

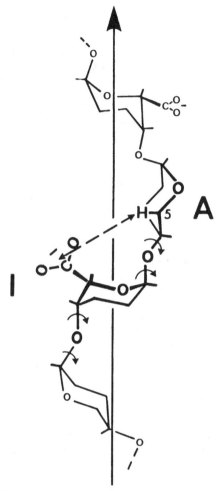

Fig. 11.   Representation of heparin as repeating sequences of iduronate and glucosamine residues, showing the proximity of H(4) of the glucosamine to the carboxyl group of a neighboring residue (Gatti et al., 1979a).

bonded" structure is also compatible with X-ray fiber patterns (Morris et al., 1980b; Atkins et al., 1980).

Similar effects of alkali are even more evident in the $^{13}$C spectrum of hyaluronate, where substantial shifts were observed for four signals [C(1) and C(2) of the glucuronate and C(3) and C(4) of the acetamidoglucose residues] compared with neutral solutions (Bociek et al., 1980).

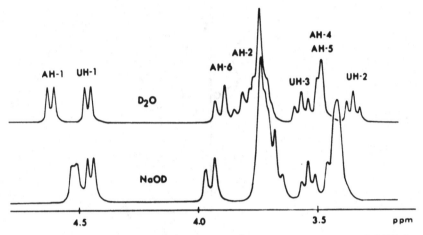

Fig. 12. Computer-simulated ${}^1$H-NMR spectra (300 MHz) for hyaluronate in $D_2O$ and in 0.4$M$ NaOD (Welti et al., 1979).

Fig. 13. Proposed "super-hydrogen-bonded" model of hyaluronate (Scott and Tigwell, 1975; Welti et al., 1979).

${}^1$H and ${}^{13}$C alkali-induced shifts are much smaller for chondroitin sulfates (Welti et al., 1979; Bociek et al., 1980).

In order to test the hydrogen-bonded model, Scott and associates are investigating the N—H and $CH_3$ signals, which are expected to be most sensitive to the local conformation of the acetamido groups and to its involvement in intramolecular H-bonding. Also O—H signals of the uronate residues provide information on their hydrogen bonding "status," as observed for monomers and dimers in dimethylsulfoxide solution (Heatley et al., 1979; Scott et al., 1981; Heatley et al., 1982).

In the 300 MHz spectrum of a tetrasaccharide from hyaluronate, in deuterated dimethylsulfoxide, the $CH_3$ signal is clearly a composite one, suggesting more than one environment for the two acetamido groups present. The N—H signals also clearly belong to two groups, one at higher field corresponding to the resonance frequency of the monomer and of N-acetylchondrosine, and the other consistently downfield (Scott et al., 1977, 1981). The N—H group giving the low-field signals exchanges with deuterium at a much lower rate than the other group, strongly suggesting its involvement in intramolecular H-bonding.

The involvement of the low-field N—H in an intramolecular association is also clearly suggested by the temperature-dependence of the corresponding signals of hyaluronate oligosaccharides. Figure 14 shows this dependence for the two types of N—H signals (denoted A and B) observed for the hexa- and the octasaccharide at fields lower than the corresponding signal of the terminal (reducing) N-acetylglucosamine residue. The signals of the oligosaccharides are much less temperature-dependent, especially at temperatures below 70°C, than the N—H signal of N-acetylchondrosine, which has the same temperature coefficient as the monomers N-acetylglucosamine and galactosamine. Since the N-acetyl groups of the disaccharide obviously cannot be involved in the intramolecular bond as proposed in Scott's model, these results provide further evidence in favor of this model, at least for hyaluronate in dimethylsulfoxide solution (Scott et al., 1977, 1981). It is worth noting that the NH signals of the octamer are at lower field relative to the corresponding ones of the hexamer. Since movement of NH (and OH) signals to lower field is usually associated with increasing strength of hydrogen bonding, it seems there is some cooperativity in the hydrogen-bond array of hyaluronate oligosaccharides.

As far as aqueous media are concerned, work by Scott and Heatley (1979) has shown that in $D_2O$ solution the $CH_3$ peak of N-acetylglucosamine monomer splits into two components in hyalobiuronic acid, and for higher oligosaccharides collapses into a single peak which, however, is composite and centered at fields lower than for the monomer. These data would suggest that at least the local conformation of the acetamido groups changes when the aminosugar is incorporated in the polymer chain.

Undoubtedly, the super-hydrogen-bonded models (Scott and Tigwell, 1975; Arnott and Mitra, this volume)

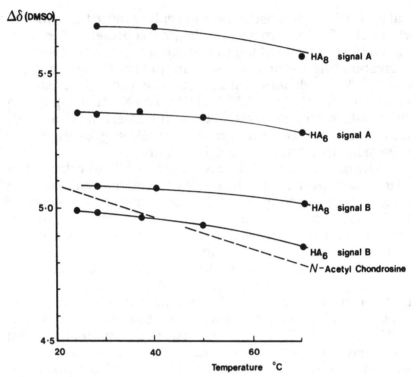

Fig. 14. Temperature-dependence of the N—H resonances of hyaluronate oligosaccharides and N-acetylchondrosine in solution in deuterated dimethylsulfoxide (300 MHz) (Scott et al., 1977).

will be extensively investigated in the future, especially with the purpose of probing by spectroscopic methods t o what extent it represents the conformation of hyaluronate chains in neutral aqueous solutions. Comparison of the conformations of hyaluronate and chondroitin sulfates in alkali is also of interest, considering the much smaller alkali-induced shifts observed in the [1]H spectra of the latter (Welti et al., 1979).

Since intramolecular hydrogen bonds conceivably affect the flexibility of the polysaccharide chain, they can be expected to influence also the NMR *relaxation times.* To this reviewer's knowledge, the only investigation along this line has been made by Rees and associates, who interpreted the non-Lorentzian line shape of the acetyl peak of hyaluronate as resulting from restricted segmental motion

(Darke et al., 1975). From these earlier measurements, it appeared that about half of hyaluronate was not contributing to the signal intensity because of pronounced line broadening by restricted motion in regions of the polysaccharide. This behavior appeared paralleled by a bimodal distribution of spin-spin relaxation times ($T_2$). However, more refined measurements of $T_2$ values favored a model in which all the segments are in continuous equilibrium between stiff and flexible states, i.e., any stiffening interactions (such as hydrogen bonds) would be continuously making and breaking (Ablett et al., 1982). Such a model is also consistent with a recent $^{13}$C-NMR study (Bociek et al., 1980).

# Chapter 5

# Self-Association of Copolymeric Glycosaminoglycans (Proteoglycans)

## L.-Å. Fransson and L. Cöster

Department of Physiological Chemistry 2,
University of Lund, Lund, Sweden

## I. A. Nieduszynski, C. F. Phelps, and J. K. Sheehan

Department of Biological Sciences,
University of Lancaster, Lancaster, UK

### Introduction

Connective tissue is composed of scattered cells embedded in an extracellular matrix consisting of abundant collagen fibers, elastin, and an amorphous ground substance dominated by proteoglycans. Proteoglycans contain a central protein core that is substituted with glycosaminoglycans. With the possible exception of hyaluronate all of the known glycosaminoglycans (Table 1) occur as proteoglycans. The various proteoglycans of the interfibrillar space are considered to control the architecture of the fibrillar network (Muir and Hardingham, 1975). A general observation is that the type of proteoglycan found in a tissue varies with the biomechanical properties of the tissue. The cartilage proteoglycan is especially designed for resisting compression and deformation by virtue of its aggregation with hyaluronate (carbohydrate–protein interaction) to form large, highly expanded supramolecular aggregates (Muir and Hardingham, 1975).

TABLE 1
Classes of Connective Tissue Glycosaminoglycans

| Homopolymers[a] | Copolymers[a] | Major building blocks[b] |
|---|---|---|
| Hyaluronate | | GlcA-GlcNAc |
| Chondroitin sulfate | | GlcA-GalNAc-$SO_4$ |
| Keratan sulfate | | Gal-GlcNAc-$SO_4$ |
| | Dermatan sulfate | GlcA-GalNAc-$SO_4$ |
| | | IdA-GalNAc-$SO_4$ |
| | | IdA-GalNAc($-SO_4$) |
| | | $\quad$ \| |
| | | $SO_4$ |
| | Heparan sulfate | GlcA-GlcNAc |
| | | GlcA-GlcNSO$_3$($-SO_4$) |
| | | IdA-GlcNSO$_3$($-SO_4$) |
| | | IdA-GlcNSO$_3$($-SO_4$) |
| | | $\quad$ \| |
| | | $SO_4$ |

[a]It should be stressed that the terms homopolymeric and copolymeric emphasize the structural features of the main core of the glycan chain. Thus, copolymeric chains contain significant proportions of two or more types of building blocks in a variety of sequential arrangements. Irregular sugar residues related to the carbohydrate–protein linkage region may be found in all the glycosaminoglycans.

[b]GlcNAc, N-acetylglucosamine; GlcNSO$_3$, N-sulfamidoglucosamine; GalNAc, N-acetylgalactosamine; UA, hexuronic acid; IdA, L-iduronic acid; GlcA, D-glucuronic acid; $-SO_4$, ester sulfate.

There are several polysaccharides of plant, microbial, or animal origin that are gel-forming, and a large variety of glycans display weak carbohydrate–carbohydrate interactions (Rees, 1972). Glycans that associate to form stable networks are usually copolymeric or have a "masked repeating structure," i.e., they are composed of more than one type of building block, whereas chains that associate more weakly are homopolymeric, i.e., they contain only one major type of building block. Mammalian glycosaminoglycans may also be divided into these categories (Table 1). In the case of hyaluronate, which is an unusually large molecule, long-range chain–chain interactions occur. These may be expressed as alterations in the rheological properties (from predominantly viscous to predominantly elastic) as a function of the strain frequency (Gibbs et al., 1968). No association phenomena involving chondroitin sulfate or keratan sulfate have hitherto been described.

The present review describes the self-associating properties of the two copolymeric variants, dermatan sulfate and heparan sulfate, and the aggregation of proteodermatan sulfate.

## Dermatan Sulfate

### Affinity Chromatography

In these experiments a homopolymeric and a copolymeric dermatan sulfate preparation (BL-18 and PS-25 in Table 2) as well as chondroitin 4-sulfate were covalently linked to agarose (Fransson, 1976). When various copolymeric chains were applied to the PS-25 gel in the presence of 0.15$M$ NaCl, the results shown in Table 3 were obtained. It is seen that PS-25 and BA-35 showed the highest degree of binding. Bound material could be displaced either by chaotropic agents (urea, guanidine) or by raising the ionic strength (0.5$M$ NaCl). It was also noted that 0.5$M$ sodium acetate displaced a portion (25%) of the bound PS-25 and all BL-18 and CS4 from the PS-25 gel. In the presence of 0.15$M$ NaCl only the strongly copolymeric preparations BA-25, PS-25, and BA-35 (Table 2) were bound to homopolymeric ligands (BL-18 and CS4, respectively). This binding was abolished by 0.5$M$ sodium acetate. The conclusions from these studies are as follows. In the presence of 0.15$M$ NaCl, homopolymeric galactosaminoglycans (i.e., those that contain essentially one type only of uronate residue) do not bind to homopolymeric ligands. However, binding between copolymeric and homopolymeric chains can be demonstrated under these conditions. In the presence of $0.5M$ sodium acetate binding between such polymers is not observed. In this solvent only copolymeric chains can interact with copolymeric chains. Apparently, this interaction is maximally expressed when the ratio of the proportions of the two types of uronate residue is close to unity or, possibly, if the species used as ligand has the same uronate composition as the free species.

It should also be added that in the present experimental design binding of free chains to one another could compete with binding of free chains to the matrix-bound ligands. Similarly, ligands with complementary binding zones might also interact, which would reduce the amount of ligand accessible for binding to free chains. These considerations lead to the following possibility. The interaction between copolymeric dermatan sulfate chains is sufficiently weak to be demonstrable as free chain–ligand interaction at a concentration of ~1 mg mL$^{-1}$ in $0.15M$ NaCl. The failure to demonstrate binding, e.g., with chondroitin sulfate onto a chondroitin sulfate gel, does not exclude the possibility that chondroitin sulfate may self-associate; the binding be-

TABLE 2
Uronic Acid Composition of Dermatan Sulfate
Preparations[a]

| Code | Source | L-Iduronate, % | D-Glucuronate, % |
|------|--------|----------------|------------------|
| BL18 | Bovine lung | 95 | 5 |
| PS18 | Porcine skin | 90 | 10 |
| PS25 | Porcine skin | 90 | 10 |
| BA25 | Bovine aorta | 85 | 15 |
| PS36 | Porcine skin | 75 | 25 |
| BA35 | Bovine aorta | 60 | 40 |
| HA50 | Human aorta | 25 | 75 |

[a]Values are given in % of total hexuronate and were calculated from carbazole-to-orcinol ratios.

TABLE 3

Results of Affinity Chromatography of Various
Dermatan Sulfates and Chondroitin Sulfates
(CS4) on Dermatan Sulfate-Agarose[a]

| Sample | BL-18 gel | PS-25 gel | CS4 gel |
|--------|-----------|-----------|---------|
| BL18   | 0         | 20        | 0       |
| BA25   | 10        | 30        | 10      |
| PS25   | 15        | 50        | 10      |
| BA35   | 20        | 50        | 20      |
| BA50   | 0         | 15        | 0       |
| CS4    | 0         | 15        | 0       |

[a]The agarose gel contained BL-18 (2 mg/mL
of gel), PS-25 (3 mg/mL of gel) or CS4 (2 mg/mL
of gel), and 1 mg of material per 0.1 mL of 0.15$M$
NaCl was applied. Elution was performed with
0.50$M$ NaCl. Values are expressed as % bound
material of total applied. No material was bound
to a blank column.

tween free chains may be so strong that it cannot be observed under the present conditions.

## Gel Chromatography

As shown in Fig. 1, the galactosaminoglycans PS-25,
BA-35, and HA-50 yielded asymmetrical peaks when chromatographed on Sepharose 6B in the presence of 0.5$M$ sodium acetate. The material was subdivided into fractions I,
II, and III. When the three preparations were chromatographed on Sepharose CL6B in the presence of 0.5$M$ guanidine hydrochloride symmetrical peaks were obtained.
These peaks had the same elution volume as did the most
retarded components obtained by chromatography in 0.5$M$
sodium acetate. In affinity chromatography experiments,
fraction I-II of PS-25 and fractions I and II of the aortic
glycans showed binding to a dermatan sulfate-substituted
gel in the presence of 0.5$M$ sodium acetate. The fractions III
were not bound under these conditions. It should be added
that the homopolymers (BL-18 and CS4) were eluted in the
same position in 0.5$M$ sodium acetate and in
0.5$M$ guanidine hydrochloride (see Fransson, 1976;
Fransson and Cöster, 1979). It may thus be concluded that

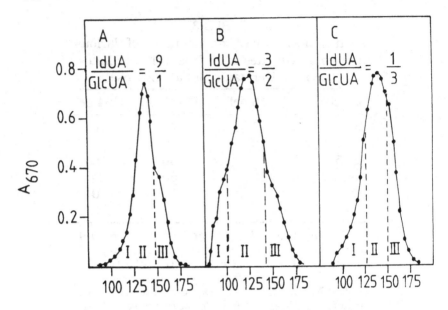

EFFLUENT  VOLUME (ml)

Fig. 1. Gel chromatography of copolymeric galactosaminoglycans on Sepharose 6B. Column size, 15 × 1400 mm;
$V_0$, 80 mL; eluent, 0.5$M$ sodium acetate, pH 7.0; elution rate, 10
mL h$^{-1}$. The samples (25–50 mg) were (A) pig skin
galactosaminoglycan, PS-25 (0–25% ethanol fraction), (B) beef
aorta galactosaminoglycan, BA-35 (25–35% ethanol fraction) and
(C) human aorta galactosaminoglycan, HA-50 (25–50% ethanol
fraction). Fractions were pooled as indicated by vertical dashed
lines. Material was recovered after precipitation with ethanol. The
uronic acid composition of the various subfractions (I–III) was
similar to that of the original material.

fractions I and II represent aggregating (binding) species
whereas fractions III are non-aggregating (nonbinding) versions. By fractionating pig skin galactosaminoglycans into
PS-18 and PS-36 (see Table 2) a largely non-aggregating
(PS-18) and an aggregating fraction (PS-36) could be obtained (Fransson et al., 1979).

### Light Scattering

The dermatan sulfate preparations PS-18 and PS-36 (Table
2) were subjected to light scattering measurements in both
0.15$M$ NaCl and 0.15$M$ KCl (Table 4). In 0.15$M$ NaCl, PS-18

yielded a weight-average particle weight of 224,000, though the particle (presumably molecular) weight was reduced to 30,000 in $0.15M$ KCl. Similarly, PS-36 had a high particle weight, 322,000 in $0.15M$ NaCl, and a low particle (presumably molecular) weight of 25,000 in $0.15M$ KCl.

A viscometric study (Fransson et al., 1979) yielded intrinsic viscosities extrapolated to zero shear of 108 mL $g^{-1}$ for PS-18 and 128 mL $g^{-1}$ for PS-36 in $0.15M$ NaCl. These high values, compared with values of 49 mL $g^{-1}$ and 35 mL $g^{-1}$ in $0.15M$ KCl, together with marked shear dependence, indicate the presence of shear-disruptible aggregates in dermatan sulfate solutions.

The behavior of the dermatan sulfates in solution was also studied as a function of ionic strength. The particle weights for PS-18 decrease with increasing ionic strength, and at I = 0.5 essentially the same value was reached as in $0.15M$ KCl. Although the particle weight for PS-36 decreased from 322,000 to 41,000 it did not attain the value of 25,000 found in $0.15M$ KCl. Light-scattering data obtained in the presence of $0.5M$ sodium acetate gave complex results. The shape of the line formed by the extrapolates to zero angle at finite concentrations was indicative of a dissociating system. At concentrations between 4 and 10 mg mL$^{-1}$ a relatively high particle weight could be extrapolated. However, in the range 1–4 mg mL$^{-1}$ the extrapolated particle weights were 37,000 (PS-18) and 25,000 (PS-36), respectively. Using the data a dissociation constant, $K_D$ of $10^{-4}M$ was calculated.

Clearly, light-scattering and viscometric measurements made at zero or low shear have detected a level of

TABLE 4

Light-Scattering Data from Aggregating and
Non-Aggregating Dermatan Sulfate

| Sample | Solvent | $\Delta n/\Delta c$, mL g$^{-1}$ | $\overline{M}_w$, daltons |
|--------|---------|------------------|------------------|
| PS18 | $0.15M$ NaCl | 0.12 | 224,000 |
|  | $0.15M$ KCl | 0.12 | 30,000 |
| PS36 | $0.15M$ NaCl | 0.12 | 322,000 |
|  | $0.15M$ KCl | 0.12 | 25,000 |

"superaggregation" that is not readily observed in gel or affinity chromatography (Fransson, 1976; Fransson and Cöster, 1979). However, the PS-36 preparation undoubtedly shows a residual self-association when the superaggregation is disrupted. This self-association, which was also noted in gel and affinity chromatography, is tentatively assigned to dimerization. The mechanism of the shear-sensitive "superaggregation" is not understood.

## Chemistry

To investigate the copolymeric structure of aggregating dermatan sulfate chains and, possibly, the structure of interacting segments of these chains, the various galactosaminoglycans were subjected to enzymic and chemical degradations. The procedures employed are outlined in Fig. 2. Copolymeric galactosaminoglycans are composed of, at least, three types of repeating units (see Fig. 2a): IdA-GalNAc-$SO_4$ (●X), IdA(-$SO_4$)-Gal NAc(-$SO_4$) (●X) and GlcA-GalNAc-$SO_4$ (OX). These units may be arranged in different ways, e.g., block regions composed of either IdA- or GlcA-containing disaccharides, or mixed regions where IdA- and GlcA-containing units may be arranged in an alternating fashion. Sulfated IdA residues are preferentially located adjacent to, or within, GlcA-containing regions. These regions of the molecule are susceptible to degradation by chondroitinase-AC or testicular hyaluronidase. Although chondroitinase-AC cleaves all GalNAc to GlcA bonds, testicular hyaluronidase requires two or more adjacent GlcA-GalNAc-$SO_4$ periods in order to be effective (Fig. 2a→b). Thus, if the copolymer contains relatively few GlcA residues, a limited fragmentation is obtained, whereas a copolymer containing a large proportion of GlcA residues is extensively degraded by this enzyme. After degradation by hyaluronidase, IdA-containing block regions are recovered as high-molecular weight fragments, though GlcA-containing block regions primarily yield tetrasaccharides (carbohydrate sequence GlcA-GalNAc-GlcA-GalNAc). Mixed regions are split into oligosaccharides of tetra- to decasaccharide size containing both IdA and GlcA. Such oligosaccharides may be further degraded by chondroitinase-AC (Fig. 2b→c). In the example shown, tetrasaccharide fragments (carbohydrate sequence GlcA-GalNAc-IdA-GalNAc and $\Delta$UA-GalNAc-IdA-GalNAc) are produced, where $\Delta$UA represents a 4,5-unsaturated uronosyl residue. Nonsulfated IdA residues in dermatan sulfate ($\bullet$ in Fig 2a) are selectively oxidized by periodate at pH 3.0 and 4°C (Fransson et al., 1974; Cöster et al., 1975). Scission of the glycan chain may subsequently be achieved by alkaline treatment (see Fig. 2a → d). Block regions composed of IdA-GalNAc-$SO_4$ periods are degraded to low-molecular weight fragments (XC), whereas GlcA- and IdA-(-$SO_4$)-containing block regions are recovered as high-molecular weight fragments. Mixed regions containing both IdA and GlcA yield oligosaccharides of varying length and composition (general carbohydrate sequence, GalNAc-(UA-GalNAc)$_n$-R where UA is either GlcA or IdA-$SO_4$, and R is the remnant of a periodate-oxidized residue). Oligosaccharides containing GlcA may be further degraded by chondroitinase-AC (Fig. 2d→e). In the examples shown, monosaccharides (GalNAc-$SO_4$), disaccharides ($\Delta$UA-GalNAc-$SO_4$) and higher oligosaccharides [carbohydrate sequence $\Delta$UA-(GalNAc-UA)$_n$-GalNAc-R] are produced.

The result of various degradations on aggregating and non-aggregating dermatan sulfate was monitored by gel chromatography. When the aggregating and non-aggregating variants of PS-25 were subjected to periodate oxidation–alkaline elimination (Fig. 2a→d), the latter species yielded a larger proportion of oligosaccharides with the carbohydrate structure GalNAc-(UA-GalNAc)$_{4-8}$-R than the former. However, the saccharide GalNAc-UA-GalNAc-R was the major degradation product in both cases. To determine their UA composition (IdA-$SO_4$ or GlcA) they were digested with chondroitinase-AC followed by gel chromatography (Fig. 2d→e). Chondroitinase-AC degradation of GalNAc-UA-GalNAc-R yielded undegraded material, $\Delta$UA-GalNAc-R and GalNAc from both species. However, the latter two were obtained in a higher yield from the aggregating species than from the non-aggregating one. Thus, the sequence -GAlNAc-IdA-GalNAc-GlcA-GalNAc-IdA-GalNAc- was more prominent in aggregating than in non-aggregating glycans. The latter chains instead contained a large proportion of IdA-$SO_4$ residues.

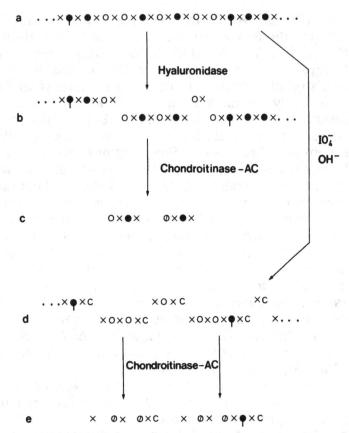

Fig. 2.   Structure and degradation of copolymeric dermatan sulfate. (a) A portion of a copolymeric chain that is composed of three types of repeating units: Ida-GalNAc-SO$_4$ (●X), IdA (-SO$_4$)-GalNAc(-SO$_4$) (♀X), and GlcA-GalNAc-SO$_4$ (Ox). Degradation with testicular hyaluronidase produces the fragments depicted in (b). A hyaluronidase-susceptible region consists of two or more adjacent GlcA-containing units (OXOX). Within this region cleavage may occur at one or two sites. Oligosaccharides containing both IdA (●) and GlcA (O) may be further degraded by chondroitinase-AC as shown in (c). Degradations with this enzyme result in the appearance of a non-reducing terminal 4,5-unsaturated uronosyl moiety (∅). Periodate oxidation of nonsulfated IdA residues (● in a) at pH 3.0 and 4°C, followed by alkaline elimination, results in a different type of fragmentation (d). Remnants of oxidized and degraded IdA moieties are present as four carbon residues at the reducing end of a GalNAc moiety (XC). GlcA (O) and IdA-SO$_4$ (♀) are not oxidized by periodate under these conditions. Oligosaccharides containing GlcA residues (O) may be further degraded by chondroitinase-AC as in (e).

Self-associating as well as non-associating variants of PS-25, BA-35, and HA-50 were digested with testicular hyaluronidase (Fig. 2a→b) followed by gel chromatography. As expected, the extent of degradation was, in general, proportional to the GlcA content of the material (see Table 2). Thus, the PS-25 variants gave a high yield of relatively large fragments, whereas the HA-50 variants yielded primarily oligosaccharides in the tetra- to decasaccharide range. The self-associating glycans afforded significantly larger quantities of medium-sized oligosaccharides, carbohydrate sequence GlcA-GalNAc-(UA-GalNAc)$_{2-4}$, than did corresponding non-associating ones. Thus, the hyaluronidase-susceptible regions of self-associating chains were often distributed in such a way that the intercalated segments were of hexa-, octa-, or decasaccharide size.

The sequences of the hyaluronidase-released oligosaccharides were also examined by chondroitinase-AC degradation (Fig. 2b→c). Oligosaccharides derived from self-associating chains yielded significant tetrasaccharide components, ΔUA-GalNAc-IdA-GalNAc, that were digestible with chondroitinase-ABC and sensitive to periodate oxidation. These results indicate that sequences such as -GlcA-GalNAc-IdA-GalNAc-GlcA- are more common in self-associating than in non-associating chains. The results summarized above were obtained both with IdA-rich and GlcA-rich copolymeric, self-associating dermatan sulfates. An alternating arrangement such as -IdA-GalNAc-GlcA-GalNAc-IdA-GalNAc-GlcA-GalNAc- was only found in self-associating chains. Non-associating chains were largely composed of block regions consisting of either IdA- or GlcA-containing repeating units. The former residues were often sulfated.

Desaccharide and higher fractions obtained from hyaluronidase digests of aggregating BA-35 contained material that was partially bound to a PS-25 gel. Furthermore, alternating carbohydrate sequences, -GlcA-GalNAc-IdA-GalNAc-GlcA-GalNAc- , were present in these saccharides. However, octa- and hexasaccharides containing alternating sequences showed no affinity for the substituted gel. A minimum chain length ($n \geq 5$) is apparently required in order to achieve sufficient cooperativity.

## Heparan Sulfate

### Gel Chromatography

Heparan sulfate from bovine lung was separated into six fractions (Table 5) of increasing charge density. The total sulfate content ranged from 0.44 to 1.63 mol/mol of hexosamine. In all fractions the O-sulfate content was the same or higher than the N-sulfate content. Estimations of the IdA and GlcA content of the various heparan sulfates by hydrolysis and chromatography indicated that fractions I–IV contained more GlcA than IdA, fraction V contained equal proportions of the two uronic acids, while fraction VI was IdA-rich. The IdA residues of heparan sulfate may be ester-sulfated, probably at C(2), rendering such a residue resistant to periodate oxidation at pH 7 and 37°C (Fransson, 1978). The proportions of IdA-SO$_4$ increased with charge density from approximately 5% of total uronic acid in heparan sulfate I to 60% in heparan sulfate VI.

Heparan sulfate III was separated into three fractions (A, B, and C) by gel chromatography on Sepharose CL6B in 0.5M sodium acetate (Fig. 3) (similar results were obtained with heparan sulfate IV). Chromatography in a dissociative medium (2M guanidine hydrochloride) yielded a sharper elution profile in the position of fraction B (Fransson et al., 1980a). Since these results suggested that the effect of guanidine may be quite complex, heparan sulfate IV was also chromatographed in different concentrations of guanidine hydrochloride (Fig. 4). In 0.5M guanidine (b) the material was eluted in a more retarded position compared with 0.5M sodium acetate (a). However, the elution profile was still irregular. When 1.0 and 4.0M guanidine were used as solvents (c and d) the peaks became symmetrical. These results were tentatively interpreted to indicate a self-association between heparan sulfate chains.

### Light Scattering

As shown in Fig. 3, heparan sulfates III (and IV) were separated into aggregatable (A) and non-aggregatable species (C) by preparative gel chromatography. The two species (A and C) were separately subjected to a light-scattering study in various solvents. The data are summarized in Table 6.

TABLE 5
Chemical Analyses of Heparan Sulfate Fractions[a]

| Fraction | NaCl, $M$ | O-sulfate/ glucosamine, mol/mol | N-sulfate/ glucosamine, mol/mol | L-Iduronic acid-O-sulfate, % | L-Iduronic acid, % | D-Glucuronic acid, % |
|---|---|---|---|---|---|---|
| I | (0.2–0.4) | 0.23 | 0.21 | 5 | 20 | 75 |
| II | (0.4–0.6) | 0.30 | 0.26 | 10 | 20 | 70 |
| III | (0.6–0.8) | 0.60 | 0.40 | 20 | 15 | 65 |
| IV | (0.8–1.0) | 0.68 | 0.47 | 25 | 15 | 60 |
| V | (1.0–1.2) | 0.61 | 0.62 | 45 | 5 | 50 |
| VI | (1.2–1.4) | 0.91 | 0.72 | 60 | 5 | 35 |

[a]Heparan sulfate was fractionated with cetylpyridinium chloride using increasing concentration of NaCl (Fractions I–IV). Total sulfate was determined after hydrolysis in $6M$ HCl at 100°C for 8 h, and N-sulfate after hydrolysis in 0.04$M$ HCl at 100°C for 1.5 h. O-Sulfate = total sulfate minus N-sulfate. Quantitation of L-iduronic acid (sulfated as well as nonsulfated) and D-glucuronic acid was accomplished by ion exchange chromatography after acid hydrolysis, deaminative cleavage, and rehydrolysis. Sulfated L-iduronic acid was determined after periodate oxidation at pH 7.0 and 37°C for 24 h. Residual carbazole-positive material was considered to represent sulfated L-iduronic acid.

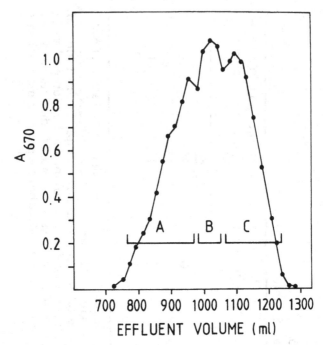

Fig. 3. Gel chromatography of heparan sulfate III. Column: Sepharose CL6B; size, 35 × 1600 mm; amount applied, 1.9 g; eluent, 0.5$M$ sodium acetate, pH 7.0; rate 20 mL h$^{-1}$; analysis, orcinol. Fractions (A–C) were pooled as indicated by vertical lines and material was recovered by ethanol precipitation followed by drying with absolute ethanol and diethylether. $V_O$ = 600 mL; $V_t$ = 1500 mL.

In 0.15$M$ NaCl or KCl the aggregatable chains yielded weight-average particle weights of 60,000–100,000, whereas the particle (presumably molecular) weight was reduced to approximately 20,000 in 4.0$M$ guanidine hydrochloride. Non-aggregatable chains afforded "monomeric" values in 0.15$M$ NaCl or KCl.

## Chemistry

Heparan sulfate molecules have a complex structure (see Fig. 5a). The principal types of repeating units include GlcA-GlcNAc (usually without ester sulfate), GlcA-GlcNSO$_3$, IdA-GlcNSO$_3$, and IdA(-SO$_4$)-GlcNSO$_3$ (in the latter three

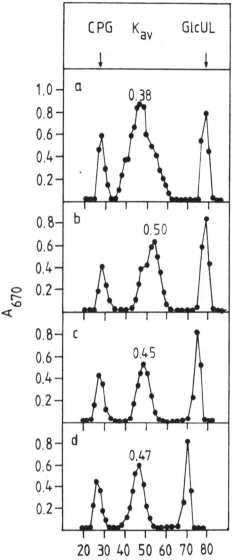

Fig. 4.   Gel chromatography of heparan sulfate IV. Column: Sepharose CL6B; size, 8 × 1400 mm; amount applied, 3 mg; eluent, (a) 0.5$M$ sodium acetate, pH 7.0; (b)–(d), respectively 0.5, 1.0, and 4.0 $M$ guanidine hydrochloride, pH 7.0; rate, 10 mL $h^{-1}$; analysis, orcinol. Heparan sulfate was co-chromatographed with cartilage proteoglycan (CPG; 1 mg) and D-glucuronolactone (GlcUL; 0.5 mg). $V_O$ = 27 mL; $V_t$ = 80 mL (a and b), 75 mL (c) and 70 mL (d). The $K_{av}$ values are indicated in the graph.

TABLE 6
Light-Scattering Data from Aggregating and
Non-Aggregating Heparan Sulfate

| Sample | Solvent | $\Delta n/\Delta c$, mL g$^{-1}$ | $\bar{M}_w$, daltons |
|---|---|---|---|
| Aggregating | 0.15$M$ NaCl | 0.12 | 63,600 |
| heparan sulfate | 0.15$M$ KCl | 0.12 | 101,600 |
| (Fraction A) | 4.0$M$ GuHCl | 0.13 | 19,700 |
| Non-aggregating | 0.15$M$ NaCl | 0.13 | 25,900 |
| heparan sulfate | 0.15$M$ KCl | 0.13 | 20,000 |
| (Fraction C) | | | |

cases with or without ester sulfate at C(6) of the amino sugar). There are two different chemical procedures available for a controlled degradation of heparan sulfates. Deaminative cleavage of GlcNSO$_3$-UA bonds by HNO$_2$ produces the fragments shown in Fig. 5b. Blocks of UA-GlcNSO$_3$ repeats are cleaved into disaccharides, and blocks of GlcA-GlcNAc are released in oligosaccharide form. In addition, segments composed of alternating UA-GlcNAc and UA-GlcNSO$_3$ repeats give rise to tetrasaccharide fragments. Such tetrasaccharides include the sequence IdA-GlcNAc-GlcA-GlcNSO$_3$ with 2,5-anhydromannose residues in terminal positions, as demonstrated previously with intermediates in the biosynthesis of heparin (Höök et al., 1974). As shown in Fig. 5c periodate oxidation at pH 3 and 4°C, followed by scission in alkali, selectively degrades the (GlcA-GlcNAc)-containing block regions into GlcNAc-R compounds (Fransson, 1978), where R represents the remnant of a periodate-oxidized glucuronate residue. The GlcNSO$_3$-containing sections of the chains are released in oligosaccharide form. Medium-sized oligosaccharides ($n = 1$–4) have the general structure GlcNAc-(GlcA-GlcNSO$_3$)$_n$-R, whereas longer oligosaccharides are highly sulfated and contain IdA residues with or without ester-sulfate (Fransson et al., 1980b). The presence of alternating arrangements of GlcNAc and GlcNSO$_3$ as well as of GlcA and IdA has been indicated.

These two complementary methods were used to determine the chemical characteristics of aggregatable (a) and

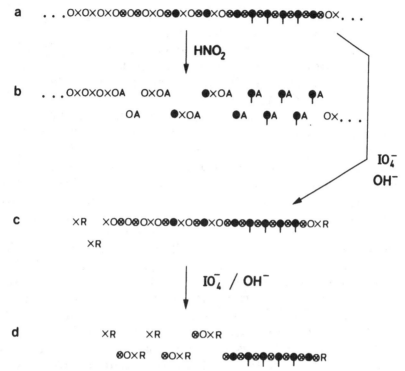

Fig. 5. Structure and degradation of copolymeric heparan sulfate. (a) The structure shown represents a portion of a copolymeric chain that is composed of several different repeating units, where GlcA (O) may be associated with GlcNAc (X) or GlcNSO$_3$ (⊗), whereas IdA (●) and IdA-SO$_4$ (●) are largely found in N-sulfated regions. The GlcNSO$_3$ moieties may also be 6-sulfated to a variable degree (not shown). Deaminative cleavage of bonds between ⊗ and O, ● or ● results in the fragmentation shown in (b); A = anhydromannose. Selective periodate oxidation of GlcA (O) in N-acetylated block regions followed by alkaline elimination releases the N-sulfated regions in the form of large oligosaccharides (c). The GlcA residues (O) of the latter saccharides are sensitive to reoxidation with periodate under the same conditions; R = remnant of oxidized and degraded GlcA.

non-aggregatable (c) heparan sulfate (Fig. 3). The results of HNO$_2$ degradation of the aggregating (a) and non-aggregating (c) versions of heparan sulfate III (HS III) and IV (HS IV) were inconclusive. The amount of (GlcA-GlcNAc)-block regions was lower in HS III-A than in HS III-C. In HS IV the situation was reversed. The yield of tetrasaccharide

(e.g., ●,XOA in Fig. 5b) was slightly higher than that of disaccharide in both the aggregating species.

Periodate oxidation–alkaline elimination of HS IV-A
and -C afforded a larger proportion of medium-sized oligosaccharides from Fig. 5a than from Fig. 5c. The larger as
well as the medium-sized oligosaccharide fractions were
pooled, and reoxidized with periodate under the same conditions. In this case, GlcA residues (associated with
GlcNSO$_3$ moieties) that survive the initial oxidation are de-
stroyed (Fig. 5c→d) and subsequent alkali treatment results in further fragmentation. Oligosaccharides derived
from aggregatable chains were extensively fragmented upon
reoxidation and cleavage. In contrast, corresponding
oligosaccharides derived from nonaggregatable chains were
affected only to a small extent. Oligosaccharides from self-
associating chains were largely fragmented into GlcN-UA-
GlcN-R and GlcN-UA-GlcN-UA-GlcN-R, whereas the corresponding saccharides from non-associating chains yielded
much larger fragments ($n = 4$ seems to be the major peak).
The UA residues of these saccharides are either IdA or
IdA-SO$_4$. Further analyses by periodate oxidation at 37°C
and pH 7 revealed that the UA of fragments from self-
associating chains were IdA. Thus the latter chains are
characterized by the presence of alternating or "mixed" sequences such as -GlcA-IdA-GlcN-GlcA-GlcN-IdA-GlcN-. This
is analogous to the situation observed with aggregating
dermatan sulfate chains.

When considering the biosynthesis of heparan sulfate
and heparin (Lindahl et al., 1977) it is evident that an
alternating arrangement of IdA- and GlcA-containing repeats can be achieved at several stages during the
biosynthesis. This process involves the following stages:

(a) Polymerization of UDP-GlcA and UDP-GlcNAc to
(GlcA-GlcNAc)$_n$
(b) N-deacetylation (partial or complete) followed by
N-sulfation
(c) C(5)-epimerization of GlcA residues located adjacent to a GlcNSO$_3$ moiety (sequence GlcNSO$_3$-GlcA-
GlcNSO$_3$/Ac) to form IdA
(d) Ester sulfation of IdA as well as GlcNSO$_3$ units

The latter process strongly promotes the net formation of
IdA. The retention of a single GlcNAc residue at stage (b)

would result in the formation of the sequences -GlcNSO$_3$-IdA-GlcNAc-GlcA-GlcNSO$_3$-. Also in (GlcA-GlcNSO$_3$)$_n$ block regions, GlcA residues may be retained if the extent of ester sulfation is low. In addition, premature C(6)-sulfation of GlcNSO$_3$ may partially inhibit C(5)-epimerization. These processes would lead to another type of sequence, e.g., -GlcA-GlcNSO$_3$-IdA-GlcNSO$_3$-GlcA-.

## Proteodermatan Sulfate

### Extraction and Purification

Fibrous connective tissues contain an abundance of collagen fibers and, in the interfibrillar matrix, proteoglycans. These proteoglycans are difficult to extract quantitatively in an undegraded state. By using extraction with 4M guanidine hydrochloride in the presence of protease inhibitors, followed by separation of collagen and proteoglycans by ion exchange chromatography in urea, a crude proteoglycan preparation from sclera was obtained in an undegraded state and in 85% yield (Cöster and Fransson, 1981; Cöster et al., 1981). The use of protease inhibitors gave highly reproducible results in contrast to preparations without. Further purification was achieved by density gradient centrifugation.

### Characterization

The purified proteoglycan preparation contained 16% hexuronic acid, 50% protein, and was free of collagen. Upon gel chromatography on Sepharose CL2B and CL4B, two polydisperse proteoglycan populations were observed, the larger of which was included on 2B but was excluded on 4B. The proteoglycan nature was demonstrated by alkali treatment and papain digestion, which yielded polysaccharide material that was eluted in the position of galactosaminoglycan chains on Sepharose CL4B.

The glycosaminoglycan chains were exclusively galactosaminoglycans. Periodate oxidation–alkaline elimination treatment of the side-chains demonstrated that all contained L-iduronic acid. The original proteoglycan must thus be classified as a dermatan sulfate proteoglycan.

## Characterization of the Proteoglycan Subfractions

By gel chromatography of Sepharose CL2B or CL4B, the proteoglycan was subdivided into two populations. The larger proteoglycan (I) represented 40% and the smaller proteoglycan (II) 60% by weight of the starting material.

To obtain molecular weights of the two proteoglycans, ultracentrifugation was carried out in $6M$ guanidine hydrochloride, which was intended to prevent any tendency to aggregate. In sedimentation velocity ultracentrifugation, both proteoglycans I and II showed the pattern of a polydisperse material, but neither was resolved into more components. Determinations of $S_{20,w}[1S = 10^{-13}s]$ and $D_{20,w}$ by ultracentrifugation for proteoglycan I (2.8 S and 1.3 $\times$ $10^{-7}$ cm$^2$ s$^{-1}$) and proteoglycan II (2.2 S and 2.0 $\times$ $10^{-7}$ cm$^2$ s$^{-1}$) were used for calculation of their molecular weights. For proteoglycans I and II, values of 170,000 and 90,000, respectively, were obtained. Meniscus-depletion sedimentation equilibrium ultracentrifugation gave similar results (I, 210,000; and II, 88,000). Thus, scleral proteoglycans are quite small compared with cartilage proteoglycans (2.5 $\times$ $10^6$) and more comparable to corneal keratan sulfate proteoglycans (72,000).

Proteoglycan I had a lower content of protein (45%), a higher content of polysaccharide material (19.5% hexuronic acid), and only 20% of the uronic acid in the galactosaminoglycan chains was IdA, whereas proteoglycan II had a higher content of protein (60%), a lower content of polysaccharide material (12.2% hexuronic acid), and approximately equal amounts of IdA and GlcA in the galactosaminoglycan chains.

Preparation of the core proteins by chondroitinase-ABC digestion of the proteoglycans followed by gel chromatography on Sepharose CL4B revealed that the core protein of I was larger than that of II. On SDS-polyacrylamide gel electrophoresis of the proteoglycan II core protein an apparrant molecular weight of 46,000 was obtained. Attempts to determine the molecular weight of the proteoglycan I core protein were unsuccessful.

Using the molecular weights obtained for the proteoglycans, its protein core, the protein content, and the molecular weight determined for the side-chains by light-scattering (see below), it was calculated that proteoglycan II carried only two side-chains whereas proteoglycan I carried five side-chains. This should be compared with cartilage proteoglycans that may carry as many as 100 chondroitin sulfate and 30–60 keratan sulfate side chains.

The dermatan sulfate side chains of proteoglycan I were largely composed of GlcA-GalNAc-SO$_4$ repeats with few IdA-GalNAc-SO$_4$ disaccharides, intercalated between short segments of GlcA-containing repeats. Although the dermatan sulfate side chains of proteoglycan II contained a large proportion of short sequences with GlcA, the intercalated IdA-containing repeats often carried an ester sulfate on the uronic acid moiety.

## Self-Association of Scleral Proteoglycans

As proteoglycan I altered its elution position upon gel chromatography on Sepharose CL4B in 4$M$ guanidine hydrochloride compared to 0.5$M$ sodium acetate, it may be proposed that a change in conformation or a disaggregation took place. The latter hypothesis was favoured as proteoglycan I appears smaller in 4$M$ guanidine hydrochloride, and after reduction and alkylation. A conformation change induced by guanidine hydrochloride or by cleavage of disulfide bonds should expand the molecule. Proteoglycan II did not show any such behaviour upon gel chromatography.

The aggregation must result from self-association, or from mediation by an extrinsic molecule, or both. Self-association may be mediated by the side chains and/or the protein core in three different ways, i.e., carbohydrate–carbohydrate, carbohydrate–protein, and protein–protein. Extrinsic molecules known to multimerize proteoglycans are hyaluronate and "link-proteins" (Muir and Hardingham, 1975). The finding that reduction and alkylation abolished aggregation might suggest that hyaluronic acid was involved, since similar results have been obtained for cartilage proteoglycans. The scleral proteoglycans were of rather low buoyant density, and therefore the possibility that small amounts of endogeneous hyaluronic acid were present in the purified material could not be excluded. Because the proteoglycans were purified by ion exchange chromatography followed by density gradient centrifugation in a dissociative medium, the presence of an aggregating protein seems rather unlikely. Since free dermatan sulfate chains from various sources may self-associate, self-association of proteoglycan I could be mediated by its dermatan sulfate side chains.

To ascertain whether multimerization could be mediated by hyaluronic acid, the purified proteoglycans were mixed with hyaluronate. The presence of high-molecular weight hyaluronic acid did not alter the elution position of the proteoglycans upon gel chromatography on Sepharose 2B. The possibility that the preparation may contain small amounts of endogenous hyaluronate was ruled out by ion exchange chromatography, which separates hyaluronic acid from proteoglycans. Upon subsequent gel chromatography, no change in elution position compared with the starting material could be detected before and after addition of hyaluronate.

It could be argued that a putative hyaluronic acid-binding region might be irreversibly denatured during chromatography on DEAE-cellulose in the presence of urea. Therefore, scleral proteoglycans were purified by two consecutive dissociative density gradients under conditions that would preserve the aggregatability of cartilage proteoglycan monomers. The gel chromatography behavior of this preparation was not affected by the addition of hyaluronate. Nor did subsequent ion exchange chromatography change the gel chromatographic behavior.

Light-scattering measurements were performed under various conditions and the data are compiled in Table 7. In $0.5M$ sodium acetate very high molecular weights were obtained. Attempts to dissociate the aggregates were made by using increasing concentrations of guanidine hydrochloride, by increasing the ionic strength, and by changing the ionic composition. In $0.15M$ NaCl the proteoglycans I and II gave particle weights of 3.1 and 3.4 $\times$ $10^6$, respectively. The aggregates would be dissociated by increasing concentrations of guanidine hydrochloride. In $6M$ guanidine hydrochloride molecular weights of 410,000 for I and 130,000 for II were calculated. These values should be compared with the values obtained by sedimentation equilibrium centrifugation (210,000 and 90,000, respectively). A possible explanation for the discrepancy is that a complete disso-

TABLE 7
Light-Scattering Data for Proteodermatan Sulfate Subfractions
(PG I and PG II) and Cognate Chains

| Solvent (all pH 7.4) | $\overline{M}_w \times 10^{-6}$ | | |
| | PG I | PG II | Dermatan sulfate chains |
| --- | --- | --- | --- |
| 0.5$M$ sodium acetate | 12 | 39 | — |
| 0.15$M$ NaCl | 3.1 | 3.4 | 0.100 |
| 0.15$M$ KCl | 3.2 | 2.8 | 0.075 |
| 1.0$M$ KCl/0.02$M$ EDTA | 2.9 | 2.3 | 0.024 |
| 6$M$ guanidine hydrochloride | 0.41 | 0.13 | — |

ciation of the aggregates was not achieved in the light-scattering experiments. Decreasing the pH from 8 to 5 in 0.15$M$ NaCl did not alter the particle weights. Additions of small amounts of hyaluronic acid or hyaluronic acid oligosaccharides had no effect on the molecular weights of the proteoglycans. These findings are in contrast with those for cartilage proteoglycans, which form aggregates with hyaluronate.

Free dermatan sulfate chains self-associate in 0.15$M$ NaCl. The dermatan sulfate side chains of scleral proteoglycans also associate under the same conditions (Table 7). The scleral proteoglycans were only slightly dissociated in the presence of 1$M$ KCl/0.02$M$ EDTA, conditions that completely dissociated the cognate dermatan sulfate chains. These results could be interpreted to indicate that multimerization was not simply an effect of chain–chain interaction. However, since multimerization would involve the interaction of several side-chains, a cooperative effect might render the proteoglycan more stable to dissociating solvents than the corresponding chains. Addition of a large excess of cognate free chains to a proteoglycan should inhibit multimerization by competing for the binding sites. Surprisingly, the results showed a great increase in the particle size of the proteoglycan, after subtraction of the intensities derived from the free chains. Free chains may have interacted with several proteoglycan monomers bridging between them to form a large network. This interpretation requires the presence of two or more binding sites on each glycan chain, which is a distinct possibility.

The relatively high molecular weights obtained for the proteoglycans in the light-scattering experiments under associative conditions should be compared with the results obtained by gel chromatography under similar conditions. In the latter case, only proteoglycan I displayed association. Furthermore, the aggregates appear to be much smaller than those seen in light-scattering experiments. This may suggest that only a small number of molecules participate in the formation of very large aggregates. However, filtration of the solutions through Millipore membranes prior to the light-scattering experiments did not change the particle size obtained. It is possible that the discrepancy between the light scattering and the gel chromatography data is caused by the presence of super aggregates that are sensitive to shear, in analogy with previous findings for free chains (Fransson et al., 1979).

The scleral dermatan sulfate proteoglycans have few side chains, and the protein core may not be constrained to an extended conformation. Therefore the conformation of the protein core may affect the aggregatability, e.g., by controlling the spatial arrangement of the side chains. Reduction and alkylation of the scleral proteoglycan I resulted in loss of aggregatability in gel chromatography experiments, whereas it was unchanged when studied by light scattering. Reduced and alkylated proteoglycan II formed aggregates that were larger than those formed by the untreated material in light-scattering experiments.

In summary, two separate dermatan sulfate proteoglycans have been isolated, with different molecular weights, protein cores, hexuronic acid contents, and carbohydrate structures of the side chains. The larger proteoglycan exhibits aggregation both in gel chromatography and light-scattering experiments, whereas the smaller proteoglycan shows aggregation only in light-scattering experiments. The multimerization of the proteoglycans is not mediated by hyaluronate or other extrinsic molecules. The remaining interpretation is that scleral proteoglycans have the ability to self-associate. We propose that the aggregation is partly mediated by interaction between the dermatan sulfate side chains. However, our data do not exclude other forms of self-association, e.g., via polysaccharide–protein and protein–protein interactions.

## Chapter 6

# Transport of Molecules in Connective Tissue Polysaccharide Solutions

## Barry N. Preston,* Torvard C. Laurent,[†] and Wayne D. Comper*

*Department of Biochemistry, Monash University, Clayton, Victoria, Australia, and †Department of Medical and Physiological Chemistry, University of Uppsala, Uppsala, Sweden

### Introduction

Connective tissue is characterized by a low cell density and a large extracellular space filled with a matrix of specialized macromolecules that determine the morphological structure and mechanical properties of the tissue. As discussed in other contributions in this volume, the family of connective tissue polysaccharides (glycosaminoglycans, mucopolysaccharides) makes up an important constituent of the matrix. These polysaccharides appear to be multifunctional (Comper and Laurent, 1978). The present review will be limited to the role that they can play in regulating transport in the tissues. A considerable amount of information has

119

been obtained from model experiments in vitro. We will discuss such model experiments although they have been performed with polymers other than the connective tissue polysaccharides.

Transport processes in connective tissues have also been dealt with in earlier reviews (Laurent, 1966; Comper and Laurent, 1978) and we will therefore concentrate the material in this review to a discussion of the principal modes of transport and the more recent developments in the field.

## The State of Polysaccharides in Connective Tissue

Most glycosaminoglycans are covalently linked to protein to form proteoglycans. The proteoglycan isolated from cartilage has been well-characterized and shown to have a molecular weight in the order of $2$–$3 \times 10^6$. Hyaluronate, which may not be covalently bound to protein, also has a molecular weight of the same order. Physical chemical characterization has shown that these large chain molecules have a very extended configuration in solution and that their hydrodynamic volumes are $10^2$–$10^3$ times larger than the volume occupied by the organic material [for references, see Comper and Laurent (1978) and contribution by Phelps to this volume].

When solutions of these high-molecular weight compounds are concentrated, the individual molecules will interact. Essentially two effects can be envisaged: (a) the coil molecules become compressed and (b) they entangle. Ogston and Preston (1979) have shown that the viscous behavior of dextran in concentrated solutions can be interpreted in terms of an initial shrinkage of the dextran molecules. A similar analysis of hyaluronate and proteoglycan leads to the same conclusion (Preston, unpublished observations). When hyaluronate and proteoglycan are chromatographed on an agarose gel in the presence of dextran, they are eluted at a higher volume than when chromatographed in buffer alone, which also is evidence for the shrinkage of the large coil molecules (Harper et al., 1982). Entanglement is known to occur in concentrated solutions of polymers, and evidence for entanglement in hyaluronate solutions comes from rheological studies and sedimentation analysis [for references, see Comper and Laurent (1978)]. At high concentrations the glycosminoglycans form a continuous network of chains in the extracellular matrix. The fine polymer network is anchored in the tissues through the interpenetration of coarse collagen filaments. The concentration of the polysaccharide network varies considerably from tissue to tissue, but it is also probable that local variations occur within the same tissues so that boundaries are formed that separate compartments of high and low polysaccharide concentrations. The mechanisms and kinetics of transport processes in polysaccharide-containing tissues are therefore difficult to analyse *in situ*. It is necessary to realize that there may be several modes of transport that can occur and different mechanisms by which the polysaccharides can regulate the transport of other compounds.

The mechanisms by which a polysaccharide matrix can influence solute transport seems to be causally related to the size of the migrating species compared to the molecular parameters (size, segmental mobility, and so on) of the network materials. Previous attempts to explain the observed effects have been made in terms of a "local" viscosity created by the polymer network, of obstacles that the polymer chain forms, and of the bulk viscosity of the solution.

## Modes of Transport

It has generally been accepted that any solute can be transported in tissues by either bulk flow or diffusion in the system. Transport by the former mechanism would be markedly retarded in a matrix, since it is acknowledged that polysaccharide networks offer a high resistance to the bulk flow of water [for references, see Comper and Laurent (1978)]. This results in any solvent flow within a homogeneous, polysaccharide-containing compartment being a slow process. Bulk flow can, however, play a role in an inhomogeneous compartment (see below).

Apart from bulk flow and diffusional processes, we have recently described (Preston et al., 1980) a new mode of solute transport in polymer solutions that may conceivably occur within biological systems. This transport, which may occur at extremely fast rates, is brought about through the formation of ordered coherent structures in solution by a microscopic convective mechanism. A detailed account of this type of transport will be discussed in a later section of this Chapter.

We will concentrate our discussion on transport in model systems. There are a number of experimentally differing systems of matrices (Fig. 1) in which transport processes have been studied and it can be envisaged that similar situations occur in vivo. We will discuss each of the systems shown in Fig. 1 with particular emphasis on diffusion processes, but before we can proceed to discuss matrix systems we must deal with diffusion in binary systems.

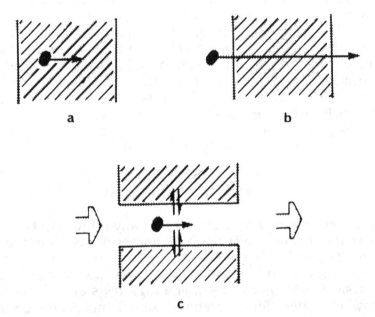

Fig. 1. Schematic representation of transport in differing matrix systems. The solvent (component 1) is shown as the stippled area, the migrating solute (component 2) as a solid particle, and the polysaccharide matrix (component 3) as the hatching. (a) Diffusive transport within an homogenous matrix; (b) diffusive transport across an homogeneous matrix; (c) transport by bulk solvent flow between matrix compartments; solute exchange between the solvent and matrix compartments occur by diffusion.

## Diffusion in a Binary System

Transport by diffusion can be characterized by two differing types of measurements yielding a mutual diffusion coefficient ($D$) or an intradiffusion coefficient ($D^*$) (tracer or "self " diffusion). As pointed out earlier (Patrick, 1961), the intradiffusion coefficient is but a special case of mutual diffusion and will be discussed as such later. The simplest system for diffusional analysis is that of a binary, solvent (component 1)–solute (component 2) system. The measurements on this system usually describe the net flux of solute across a boundary. This flux represents an exchange of solute with solvent as a result of the relaxation of the chemical potential gradient of solute. For systems with zero volume across the boundary the *mutual diffusion* coefficient [see, for example, Sundelöf (1979)] of component 2 at constant temperature ($T$) and pressure ($P$) is given by:

$$D(c) = [c/f(c)] \ (\partial\mu/\partial c)_{T,P} \qquad (1)$$

where $\mu$ is the chemical potential, $c$ the concentration of component 2, and $f(c)$ is a frictional factor. Thus there are two factors that determine the diffusion. They are usually referred to as "*the hydrodynamic factor*" $1/f(c)$ and the "*thermodynamic factor*" $[c(\partial\mu/\partial c)]$. The hydrodynamic factor describes the frictional resistance offered by the solvent and neighboring solute molecules. The thermodynamic factor describes the interaction between solute molecules, e.g., exclusion, that leads to a change in the chemical activity. Both factors are therefore functions of the solute concentration. Although this separation of $D$ into two factors can be of value in qualitative discussions, it should not be overemphasised since thermodynamic and hydrodynamic effects are closely interconnected.

The thermodynamic factor can be evaluated by standard techniques, e.g., osmotic pressure measurements, since:

$$c(\partial\mu/\partial c)_{T,P} = RT \ (1 - \bar{v}c) \ (1 + 2A_2Mc + 3A_3Mc^2 + \ldots)(2)$$

where $M$ and $\bar{v}$ are the solute molar mass and partial specific volume, respectively, and $A_2$ and $A_3$ are the standard form of the virial coefficients of component 2 (Yamakawa,

1971; Laurent et al., 1976; Kitchen et al., 1976). If we simplify the notation by identifying:

$$Q(c) = (1 - \bar{v}c)(1 + 2A_2Mc + 3A_3Mc^2 + \ldots)$$

we have from Eqs. (1) and (2):

$$D(c) = RT\, Q(c)/f(c) \tag{3}$$

In the ideal case, as $c \to 0$, Eq. (3) reduces to:

$$D_o = RT/f_o \tag{4}$$

The diffusion of the solute will only be determined by the friction with the solvent in the ideal case. According to Stokes' (1856) law, generalized by later authors, $f_o$ can be related to the solvent viscosity $(\eta_o)$ by

$$f_o = 6\pi\eta_o r \tag{5}$$

where $r$ is the radius of the "equivalent hydrodynamic sphere."

From Eqs. (3) and (4) we can express the reduced diffusion coefficient as:

$$D(c)/D_o = [f_o/f(c)]Q(c) \tag{6}$$

Clearly the interdependence of $Q(c)$ and $f(c)$ makes it difficult to evaluate $f(c)$; however, an alternative method is available through the measurement of the *intradiffusion* (self-diffusion) coefficient $(D^*)$. In the case of intradiffusion, no net flux of diffusant occurs, and the molecules undergo an exchange process. However, the molecular movement of the solute will be hindered by the solute–solvent interactions. The measurement is usually carried out by using trace amounts of labeled component 2 in a homogeneous system of component 2, free of any gradients in chemical potential. Hence $D^*$ is given by:

$$D^*(c) = RT/f(c) \tag{7}$$

or

$$D^*(c)/D_o = f_o/f(c) \tag{8}$$

Experimentation in such a system is difficult since the tracer boundary lacks the stabilizing influence of a concentration gradient. This can be overcome by the use of measurements at different concentration gradients followed by extrapolation to zero gradient (Laurent et al., 1976).

Note the frictional factor $f(c)$ in Eqs. (3) and (7) have commonly been assumed to be identical although recent studies on polymer diffusion would suggest that this is not the case (Comper et al., 1982).

### Diffusional Transport Within a Homogeneous Polymer Matrix

When a solute molecule (component 2) is moving (Fig. 1a) in a matrix (component 3) it will encounter frictional resistance from the solvent, the neighboring solute molecules, and the matrix. Similarly the thermodynamic factor in the diffusion equation will be governed by the concentrations of both components 2 and 3. The system is thus much more complex than a binary system. We will simplify our approach to the problem by regarding component 2 as moving in a medium made up by the solvent (component 1) and the matrix (component 3). The mutual diffusion coefficient of component 2 is in analogy with Eq. (2):

$$D_t(c_2,c_3) = RT \ Q_t(c_2,c_3)/f_t(c_2,c_3) \qquad (9)$$

where $c_2$ and $c_3$ now refer to the concentrations of components 2 and 3. The subscript $t$ is used to show that the frictional and thermodynamic factors refer to a ternary system.

In certain simplifying conditions we may be able to evaluate $Q_t$ and $f_t$ as independent factors. However, owing to the complex interdependence of $Q_t$ and $f_t$ in the general case, it may be more profitable to consider the behavior of the ratio $(Q_t/f_t)$ as a measure of solute behavior in matrices. The factor $(Q_t/f_t)$ will be discussed in empirical terms later.

Finally, an "intradiffusion" coefficient of component 2 can be defined, similar to that in a binary system:

$$D_t^*(c_2,c_3) = RT/f_t(c_2,c_3) \qquad (10)$$

The concentration of solute and matrix should be homogeneous throughout the system. Such an intradiffusion coefficient can be used to determine the frictional interaction between the solute and the matrix if the solute concentration is low $(c_2 \rightarrow 0)$.

From Eqs. (4) and (10) we get:

$$D^*_t(c_3)/D_o = f_o/f_t (c_3) \tag{11}$$

It is interesting that transport through polysaccharide matrices has been determined by a method other than diffusion, i.e., sedimentation. The frictional resistance in sedimentation should be similar to that in diffusion, while the driving force should not be influenced by a thermodynamic factor. Data from sedimentation analyses should therefore resemble those obtained by Eq. (11).

## Transport Across a Matrix (or Membrane)

This diffusional transport occurs by mechanisms similar to those discussed above for transport of solutes within a matrix. In experimental terms, the transport from an obstacle-free environment across a matrix into another obstacle free medium (Fig. 1b) is measured. If the obstacles under consideration are only those inherent in the matrix as discussed above, then the diffusion coefficient across the matrix is the same as defined in Eq. (9). The total flux over the membrane is, however, also determined by the partition of the solute between the obstacle-free medium and the matrix, and it is common to define the transport by a permeability coefficient ($P$):

$$P = KD_t \tag{12}$$

where $K$ is the partition coefficient of component 2 between the matrix and the solvent.

It should be noted that if the matrix–membrane is stabilized by an inert support, which is a common experimental design, then the flux of material, and hence $P$, is reduced further. It is customary in this case to allow for the obstructions offered by the supporting system by calibrating the system with a suitable solute. All subsequent measurements are given relative to the behavior of the calibrant material.

## Transport in an Inhomogeneous Matrix

This is depicted in Fig. 1c and is considered to consist of cross-linked regions of matrix interdisposed with solvent phases. Normally, the whole system is stabilized by a mechanical support. Transport can then occur by both diffusion and bulk solvent flow. The diffusional transport will be a complex function of the concentration of the matrix component and the distribution of the solute between the matrix and the surrounding solvent, as outlined above.

When solvent flow occurs in the system, the flow will essentially be between the matrix granules since the flow resistance is very high within the polymer network. Material between the matrix compartments will move, and material within the matrix compartments may be regarded as stationary. The rate of transport of material will, therefore, be dependent on the rate of solvent flow and the partition of the compound between the mobile and stationary phases. The process is comparable to a chromatographic process. If the transported species is sterically excluded from the matrix, then the process will be similar to gel chromatography. Large molecular size material will be transported more rapidly than low molecular weight material. If the moving species is adsorbed to the matrix its transport will be retarded, and the process is similar to adsorption chromatography.

Transport by gel chromatography on hyaluronate gels has been discussed in a previous review (Laurent, 1966) and no recent developments have occurred. We will, therefore, omit discussion of this mode of transport in this communication.

### Interpretation of Translational Transport Data

Interpretation of solute transport within or across networks has been attempted in terms of two alternative models of the diffusion process: (1) the "stochastic" or random walk model (Einstein, 1905), which regards diffusion as resulting from a succession of unit displacements or steps of individual particles, statistically independent of each other with regard to direction; (2) the phenomenological model (originated by Einstein, 1908) and expressed in Eqs. (1)–(11). Some limited success has been achieved by use of the stochastic model to explain the retardation of globular macromolecular solutes in aqueous solutions of chain polymers (Ogston et al., 1973). However, this approach was not successful in describing the transport of low molecular weight solutes (Kitchen, 1975) or of large polystyrene particles (Hallet and Gray, 1974; Turner and Hallet, 1976) in polysaccharide media.

Alternative approaches by use of the phenomenological model (Pappenheimer et al., 1951; Wang, 1954; McLennan, 1956) have met with varying degrees of success. However, no satisfactory general explanation has been offered that covers the transport behavior of small and large particles in various networks made up of polymers of differing molecular weight.

The lack of a satisfactory uniform approach to this problem may arise because of the inherent nature of solutions of flexible polymers. It is generally agreed that polymer solutions display critical concentrations (Aharoni, 1978) at which a measured property changes its concentration dependence. This leads to the concept of the existence of a *dilute range* in which each polymer molecule behaves independently of any other with no measurable intermolecular interactions. As the concentration $c$ is increased, the distance between the coils decreases and, at some concentration $c^*$, the coils overlap and begin to interpenetrate and form a network. This describes the beginning of the *semidilute* range. The behavior of polymers in the semidilute range has been the subject of extensive investigation [for a review, see De Gennes (1979)] and is beyond the scope of this chapter. However, recently Langevin and Rondelez (1978) using scaling techniques (Des Cloizeaux, 1975) have offered a generalized explanation for the transport behavior of solutes in semidilute and concentrated polymer solutions, in which the solution is viewed as a transient statistical network of mesh $\delta$. The retardation of the solute is considered to be governed by the elastic distortion of the network caused by the movement of the solute particles. The frictional force experienced by a spherical particle of radius "$r$" will be different according to whether "$r$" is greater or smaller than $\delta$. According to these workers, the frictional coefficient, $f(c)$, at a polymer concentration of $c$ is given by $f_0/f(c) = \psi (r/\delta)$, where $f_0$ is the frictional coefficient in pure solvent. The detailed form of $\psi$ is not predicted, but certain features are evident: (i) if $r < \delta$ then $\psi (r/\delta) \rightarrow 1$, that is the particle experiences a frictional coefficient given by Eq. (5), $f_0 = 6\pi\eta_0 r$; (ii) if $r \gg \delta$ then $\psi(r/\delta) \rightarrow \eta_0/\eta_M$ ( $= \eta_{rel}^{-1}$) where $\eta_M$ is the macroscopic viscosity of the solution, that is, the frictional coefficient is given by $f = 6\pi\eta_M r$. Since $\delta \propto c^{-0.75}$, it is suggested that:

$$D_t^*(c_3)/D_o \propto \exp(-rc^{-0.75})  \tag{13}$$

Since $\delta$ is predicted to be independent of molecular weight of the network forming species, $D_t^*(c_3)/D_o$ should also be independent of molecular weight of component 3.

It is of interest to investigate to what degree the diffusional behavior of solutes within and across matrices is in agreement with these predictions.

## Diffusion of Low Molecular Weight Solutes in Polymer Matrices—Concept of Microviscosity

From early work on the transport of low molecular weight solutes in polymer solutions (Edelson and Fuoss, 1950; Nishijima and Oster, 1956), it became evident that the motion of these particles was much less sensitive to the macroscopic viscosity than one should expect if the frictional coefficient followed the form predicted by Eq. (5). This led to the concept of a local viscosity (or microviscosity) $\eta^\mu$, describing the microenvironment seen by a diffusing particle in a polymer solution of concentration $c$. The microviscosity of solvent $\eta^\mu_{rel} = \eta^\mu/\eta_o$ is given by the ratio of the apparent frictional coefficient that the diffusant encounters in the matrix, $f(c)$, compared to that in the solvent, $f_o$ i.e.,

$$\eta^\mu_{rel} = f_t(c)/f_o = D_o/D_t^*(c)  \tag{14}$$

We have undertaken extensive measurements of the diffusion of radioactively labeled small solutes in dextran, as a function of both polymer concentration and molecular weight (Preston, Checkley, and Kitchen, unpublished observations). We have chosen to use dextran as a model compound in view of its ease of supply as fractionated material. Dextrans of various molar mass ($10^4 < \bar{M}_w < 2 \times 10$ 26) were all found to have marked effects in reducing the apparent diffusion coefficient of a number of small solutes (Fig. 2). It is evident that the size of the retardation is dependent upon the size of diffusant, but independent of the molecular size of the dextrans.

We have also carried out distribution studies for a limited number of small solutes between dextran solutions and water (or dilute salt) and these are shown for the solute glu-

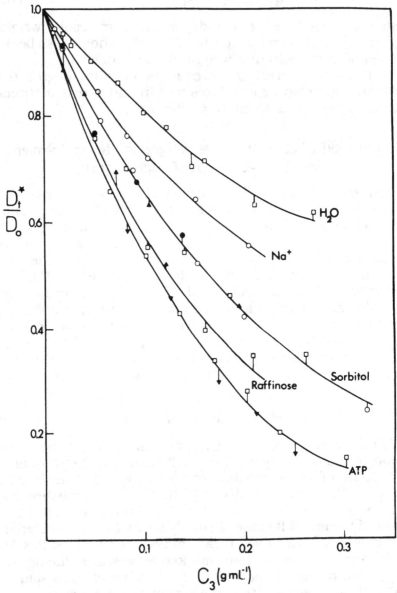

Fig. 2. The variation of the reduced diffusion coefficient.
$D_t/D_o$, of several low molecular weight solutes with change in concentration of dextran ($c_3$). Dextrans of various molecular weights were used: $M_w \sim 10^4(\square)$; $2 \times 10^4(\bigcirc)$; $7 \times 10^4(\bullet)$; $1.5 \times 10^5(\blacktriangle)$; $2 \times 10^6(\blacktriangledown)$.

cose, in Fig. 3. The partition coefficient, $K_{av}$, describes the available volume fraction for glucose in the dextran solution. It is evident from Fig. 3 that $K_{av}$ closely follows the reduction in diffusion coefficient. Our interpretation of this data suggests that the reduction in the diffusion coefficient is adequately explained by the reduction in the volume within the matrix available to the diffusant. We do not offer any quantitative explanation of why $f_o/f(c)$ should be equated to the value of $K_{av}$. We have shown that the various excluded volume treatments of Wang (1954), McLennan (1956; 1957), Ogston (1958), and Phillips et al. (1977) explain the data only if certain model-based assumptions are made.

What of the concept of microviscosity? As stated in Robinson and Stokes (1959), "this situation (of reduced mobility) is sometimes described by saying that the 'microscopic viscosity' of the solution is lower than the measured

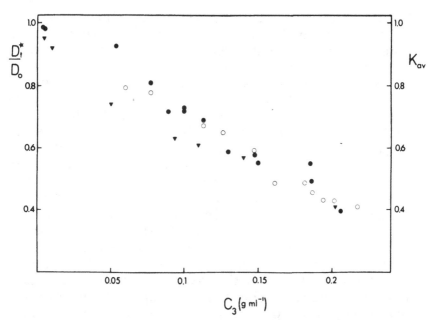

Fig. 3. Comparison of the reduced diffusion coefficient, $D_t^*/D_{o'}$ of glucose in solutions of dextran ($c_3$), $\bar{M}_w \sim 10^4$ ($\triangle$) and $\bar{M}_w \sim 5 \times 10^5$ ($\blacktriangledown$) and the partition coefficient ($K_{av}$) of glucose equilibrated between solutions of dextran ($\bar{M}_w \sim 2 \times 10^4$) and water (O).

viscosity, but this statement does not constitute an explanation of the effect."

It may be that the increased macroscopic viscosity of the solution and the increased resistance to the mobility of the particle are not related as cause and effect, but merely reflect a common cause. We believe that the microviscosity concept is but a measure of the excluded volume interactions between the diffusant and matrix. This can more clearly be seen by the following analysis. By analogy with the intrinsic viscosity of the polymer solution, $[\eta]$ defined by:

$$(\eta/\eta_o - 1)/c = [\eta] + k'[\eta]^2 c$$

where $\eta$ is solution viscosity, $\eta_o$ the viscosity of solvent, and $k'$ is the Huggins constant, we can identify an intrinsic microscopic viscosity of the polymer solution as:

$$[\eta^\mu] = \{[D_0/D_t^*(c_3) - 1]/c_3\}c_3 \rightarrow 0 \qquad (15)$$

Plots of $[D_0/D_t^*(c_3) - 1]/c_3$ against $c_3$ are shown in Fig. 4 for various low molecular weight solutes. It is interesting to note that extrapolation of data to $c_3 \rightarrow 0$ yields estimates of $[\eta^\mu]$ between 3.0 and 3.8 mL g$^{-1}$.

Similar transport data on the behavior of small solutes in aqueous solutions of other polymers such as polyethylene glycol (Elworthy et al., 1972; Phillips et al., 1977), polyvinyl alcohol (Komiyama and Fuoss, 1972), polyvinylpyrrolidone (Elworthy et al., 1972), and Ficoll (Stokes and Weeks, 1964) have been similarly evaluated. Although the polymers ranged in molecular weight from $6 \times 10^3$ to $3.6 \times 10^5$ and varied considerably in chemical composition, estimates of $[\eta^\mu]$ obtained were $\sim 3.4$ mL g$^{-1}$. The insensitivity of this parameter to molecular weight and chemical composition of the matrix material is understood in terms that $[\eta^\mu]$ is a measure of the exclusion interactions of the small diffusant molecule by segments of the linear chains of the matrix. A small expected dependency of $[\eta^\mu]$ upon the molecular size of the diffusant is observed (Fig. 4). It should be emphasized that the observed effects on the mobility of Na$^+$ (Fig. 2) arise from excluded volume phenomena as distinct from charge effects, which will be considered in a later section. The results of a recent

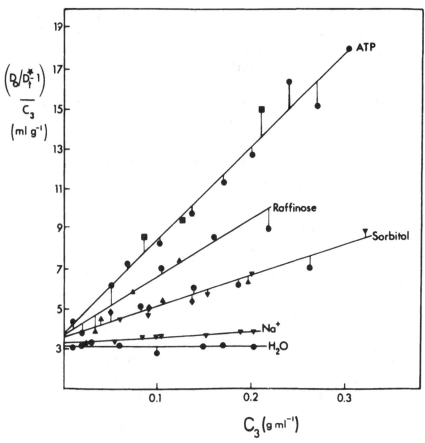

Fig. 4. Dependence of the reduced "microscopic" viscosity $(\eta^{\mu}_{rel} - 1)/c_3 = (D_O/D_t^* - 1)/c_3$ of dextran solutions of concentration $c_3$, as estimated from the diffusion of several low molecular weight solutes. Various dextrans of differing molecular weights were used: $\bar{M}_w \sim 10^4$ ($\bullet$); $2 \times 10^4$ ($\blacktriangledown$); $7 \times 10^4$ ($\blacklozenge$); $1.5 \times 10^5$ ($\blacktriangle$); $2 \times 10^6$ ($\blacksquare$).

study involving the use of paramagnetic resonance techniques to investigate the behavior of water in dextran solutions (Tsitsishvili et al., 1979), though showing similar trends to those described above (Fig. 2), were in quantitative disagreement. We can offer no explanation for this discrepancy. No comprehensive data is yet available with regard to the diffusion of small solutes in glycosaminoglycan solutions, but some recent reports by Napier and Hadler

(1978) and Hadler (1980) on the behavior of glucose in complex media containing hyaluronate are noteworthy. These workers report that the diffusion of glucose (and other small solutes) is enhanced in such media.

## Mutual Diffusion of Macromolecular Solutes in a Binary System

Some insight into the diffusional behavior of macromolecular solutes within and across matrices may be obtained by considering the simpler case of a binary solvent–solute system.

We have undertaken extensive measurements on the diffusive and other physicochemical parameters of a number of macromolecular solutes (Kitchen, 1975). Some of the data on these systems have been published elsewhere (Laurent et al., 1976; Kitchen et al., 1976; Ogston and Preston, 1979; Preston et al., 1982). As a typical example of these systems, we shall consider the behavior of a dextran sample (molecular weight $\simeq 1.5 \times 10^5$). The concentration dependence of the mutual diffusion coefficient ($D$) and of the intradiffusion coefficient ($D^*$) (measured as a $^3$H-labeled species) are shown in Fig. 5 (Preston et al., 1982). The observed behavior is typical of linear chain polymers (Roots et al., 1979).

A plot of the calculated values of $D(c)/D_o Q(c)$ is seen to correspond closely to the dependence of the intradiffusion coefficient. This agrees with the assumption inherent in Eqs. (5)–(8) that the frictional interactions are similar in both types of diffusion (see however, Comper et al., 1982).

It was suggested earlier (Bearman, 1961) that the mutual diffusion coefficient and the viscosity of the system $\eta$ are related by:

$$D(c)\eta(c) = D_o \eta_o Q(c) \qquad (16)$$

However, it has been pointed out (Vasenin and Chalykh, 1966) that this relationship does not appear to hold over the entire concentration range, and reasons for this breakdown have been offered. It is of interest to note, however, that in this system, as for many others we have studied, the equation seems to describe the diffusional process up to a

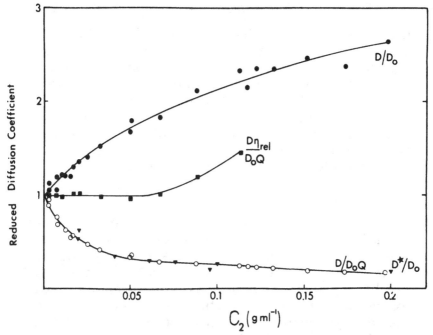

Fig. 5. Concentration dependence of the reduced mutual diffusion coefficient $(D/D_o)$ (●) and of the reduced intradiffusion coefficient $(D^*/D_o)$ (▼) of a dextran, molecular weight $(\overline{M}_w) = 1.52 \times 10^5$. Included in the plot are values of $D/D_oQ$ [as per Eq. (6)] (○) and $D\eta_{rel}/D_oQ$ [as per Eq. (16)] (■). It is evident that the dependence of $D^*/D_o$ and of $D/D_oQ$ upon the dextran concentration $(c_2)$ may be regarded as coincident.

concentration of~7% (Fig. 5). The extent to which observations made on these model systems can be used to infer the diffusional behavior of the polysaccharide components of connective tissue is uncertain. Although the polysaccharides exist at comparable concentrations (1–10%), there is extensive reasoning to suggest that the concentration level at which Eq. (16) ceases to describe adequately the diffusional process is related to the polymer molecular weight. In connective tissues this is thought to be very large, and hence the validity of Eq. (16) may be restricted to a very low concentration range.

## Diffusion of Macromolecular Solutes in Matrices

It is evident from Figs. 2–4 that molecules of different size encounter different frictional resistance in concentrated polymer solutions. Such observations had led to the concept of microviscosity but, as pointed out earlier, it may be more appropriate to discuss the increased frictional resistance in terms of a mechanism that considers the excluded volume interactions between the diffusant and matrix. Such a mechanism must allow for the dependence of the frictional interaction on the size of diffusant.

### The Obstacle Hypothesis

It has been shown that the concept of obstacles being formed by the matrix, which restrict the movement of other solutes, offers a quantitative explanation for systems where proteins are diffusants and higher molecular weight polysaccharides such as hyaluronate form the matrix. The first extensive studies on translational transport in connective tissue polysaccharide matrices were made by sedimentation. Laurent et al. (1963) carried out sedimentation analysis on various spherical proteins, a virus, a silica colloid, and several polystyrene particles in the presence of increasing concentrations of high molecular weight hyaluronate. The particle diameter of the tested substances varied between 4.7 and 365 nm. The authors observed a marked decrease in sedimentation rate with increasing hyaluronate concentration, and the relative decrease was larger the larger the size of the migrating particles. They interpreted the results as if the hyaluronate network acted as a sieve allowing small particles to penetrate while retarding large particles. The data allowed the formulation of an empirical equation of the type:

$$S/S_o = A \exp\left(-1.4rl^{1/2}\right) \tag{17}$$

where $S/S_o$ is the relative decrease in sedimentation rate compared to its sedimentation in hyaluronate-free solution, $r$ is the radius of the sedimenting particle, $l$ is the concentration of hyaluronate expressed as cm chain per $cm^3$, and $A$ is a constant, close to unity, that varies from particle to particle. Subsequent studies have shown the relationship to be valid for sedimentation in other polymer networks, and that the diffusion of globular proteins follows a closely similar dependency (Laurent et al., 1963; Laurent and Persson, 1964; Preston and Snowden, 1973a). A theoretical explanation of this behavior was offered by Ogston et al. (1973) in which they treated solute migration as a stochastic process. In this treatment, the random step movement of the solute is prevented by collision with an obstacle. The resultant retardation is then given by the probability of such a collision taking place. This probability is calculated from an earlier relationship by Ogston (1958) that described the volume available to a spherical particle in a random network of long rods. The relationship derived by these authors was:

$$D/D_o = \exp\left(-\pi^{1/2} r l^{1/2}\right) \tag{18}$$

which is in close agreement with the empirical relationship shown above.

It is interesting to note the Donners (1977) has shown that the migration of micelles through a dextran network is satisfactorily described by Eq. (18).

In logical continuation of these early studies, Laurent et al. (1975) investigated the transport of asymmetric particles in hyaluronate networks. The diffusion of various linear polymers, including DNA, was measured and it was observed in all cases that the relationship derived for spherical particles was obeyed, except that the molecular radius estimated from diffusion in the matrix was smaller than the hydrodynamic radius in free solution. The authors concluded that the asymmetric particles may have moved end-on in the network since this movement should encounter less obstacles than a random movement of the whole hydrodynamic unit. That a linear chain molecule could undergo such a movement (reptation) had been earlier suggested by De Gennes (1971).

In an investigation of the diffusion of various macromolecules across a membrane consisting of a cartilage matrix deposited in tissue culture, Cumming et al. (1979) have shown that globular proteins were retarded as expected, but that two asymmetric molecules, fibrinogen and collagen, were transported faster than their diffusional behavior in free solution would predict. Such effects could be explained if the molecules were forced into end-on movements since a rod-like particle would encounter less resistance moving lengthwise than when moving broadside.

In terms of our earlier analysis, it is obvious that Eq. (18) represents a mechanistic interpretation of the factor $(Q_i/f_i)$ of Eq. (9). A closer examination of the obstacle model of Ogston suggests that it is applicable only within certain limits with respect to molecular size and concentration of network component and to the molecular size of diffusant.

## Limitations of the Ogston Obstacle Model

The obstacle theory assumes the presence of a random network of fibers, and that the mobility of the network is effectively zero with respect to the movement of the diffusant. The criteria of network formation imply that the concentration of matrix material must be in the semidilute range, as defined earlier; below these concentrations the concept of a network is not applicable. Furthermore, it may be that at higher network concentrations the possibility of a multiple number of fibers being involved in a single collision must be considered. The limitations regarding the relative mobility of the network are intimately linked to the molecular weight of the matrix-forming material. It is obvious that a small molecule cannot form an effective obstacle in terms of the model. A further restriction must be recognized in terms of an upper limit to the size of the diffusant. If the migrating species is large compared to the molecular dimensions of network mesh, then the simple consideration of the collision involving a single fiber is inappropriate.

The dependence of solute retardation upon the molecular weight of the matrix material was shown by Laurent and Persson (1964) in sedimentation analysis. These workers studied the sedimentation of two globular proteins, α-crystallin and serum albumin, in the presence of several polymers that varied with respect to chemical composition, molecular weight, and charge. Linear charged polysaccharides of high molecular weight offered the highest resistance towards the sedimentation of the proteins. When the molecular weight of the polymers was decreased, their effect on the sedimentation of the proteins also decreased, a result which cannot be accommodated by the obstacle hypothesis that predicts the effect to be only a function of the concentration of the matrix polymer.

As pointed out earlier, the scaling treatment of Langevin and Rondelez (1978) also predicts that the retardation should be independent of the molecular weight of the matrix material.

Although recognizing the suitability of the theoretical treatment of Ogston et al. (1973) to deal with the restricted case, which seems to cover adequately consideration of protein transport in connective tissues, there is an absence of any comprehensive approach based upon restriction caused by obstacles. Therefore, it is of interest to consider the limitations of using the macroscopic viscosity of the network-forming solutions as a measure of the frictional interactions involved.

## Macroscopic Viscosity

We have recently undertaken a detailed examination of the various factors involved in solute retardation, using albumin and thyroglobulin as diffusants in dextran solutions of various molar mass ($10^4 < \overline{M}_w < 2 \times 10^6$). From the results with albumin (Fig. 6), it is evident that the reduction in the diffusion rate of the protein is dependent upon the nature of the network component with respect to both its concentration and molecular weight, in agreement with the earlier sedimentation results of Laurent and Persson (1964). However, it is seen that the retardation becomes less sensitive to the molecular size of the network material as this parameter increases. It is interesting to note that the reduction in diffusion rate of the protein in solutions of low molecular weight solutes (dextrans of $\overline{M}_w < 10^4$; sucrose) is quantitatively explained by the increased macroscopic viscosity of the solution [use of Eq. (16) with $Q(c) \rightarrow 1$ as $c_2 \rightarrow 0$ (Fig. 6)]. Similar experimental trends were observed with thyroglobulin (molecular weight $6.7 \times 10^5$) as diffusant; in this case "ideal" behavior, described by $D/D_o = (\eta_{rel})^{-1}$, was observed in dextran solutions of $\overline{M}_w < 7 \times 10^4$. An examination of the sedimentation analysis of $\alpha$-crystallin (molecular weight $8.3 \times 10^5$) in various dextran solutions (Laurent and Persson, 1964) suggests that the solution macroscopic viscosity would explain the observed reduction in sedimentation rate for matrix molecular weight $< 5 \times 10^5$. The continuing trend has been observed by Hallet and Gray (1974) and Turner and Hallet

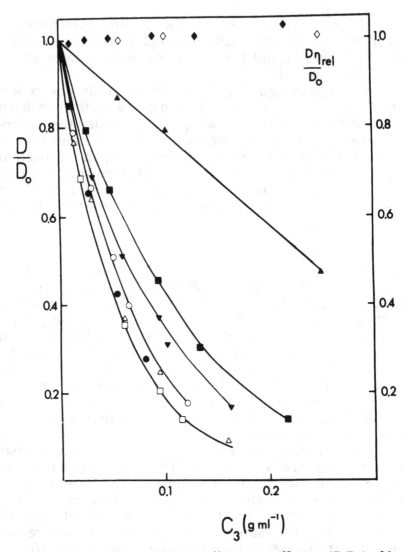

Fig. 6. Reduced mutual diffusion coefficient ($D/D_o$) of human serum albumin (HSA) in solutions of dextran (concentration, $c_3$) of various molecular weights: $\overline{M}_w \sim 10^4$ (■); $2 \times 10^4$ (▼); $7 \times 10^4$ (○); $1.5 \times 10^5$ (△); $5 \times 10^5$ (●); $2 \times 10^6$ (□). For comparison, the diffusion of HSA in solutions of sucrose is included (▲). Values of $D\eta_{rel}/D_o$ for diffusion of HSA in solutions of sucrose (◇) and of dextran, $\overline{M}_w \sim 10^4$, (◆) are shown; it is·obvious that for $c_2 \lesssim 0.2$ g mL$^{-1}$, $D\eta_{rel}/D_o$ is a constant equal to unity. All measurements were carried out in phosphate buffered saline, pH 7.0, at 20°C.

(1976), who measured the diffusion of large polystyrene particles in the presence of hyaluronate and of various dextrans. They suggested that in all cases studied the reduction of the transport rate of the polystyrene spheres was a function of the macroscopic viscosity of the solution.

An examination of the diffusion of various solutes in the presence of dextrans of varying molecular weights demonstrates where the macroscopic viscosity of the matrix ceases to be a quantitative index of the frictional forces involved in the solute migration (Fig. 7). The diffusion of $[^3H]$-$H_2O$, sorbitol, raffinose, serum albumin, and thyroglobulin is in each case dependent on the macroscopic viscosity of the system in low molecular weight dextran (or saccharides), but at a certain molecular weight of the dextran the transport tends to become independent of the macroscopic viscosity. The break-point occurs at increasingly higher molecular weights of dextran as the molecular size of diffusant is increased. It is apparent that an important factor in determining the mechanism causing the retardation of the solute is the relative size of the diffusant to that of the matrix component.

In this regard a recent proposal of a viscosity model for gel electrophoresis (Bode, 1979) may offer an alternative analysis of restricted transport.

## Summarizing View

In view of the apparent complex nature of solute diffusion within polymeric matrices it is perhaps helpful to offer some summarizing statements.

(a) LOW MOLECULAR WEIGHT SOLUTES    From the extensive published data (Edelson and Fuoss, 1950; Nishijima and Oster, 1956; Stokes and Weeks, 1964; Elworthy et al., 1972; Komiyama and Fuoss, 1972; Phillips et al., 1977), it seems that the reduction in the diffusion rate of low moelcular weight solutes (molecular weight $< 700$) in polymer solutions is consistent with transport through a medium of low viscosity, comparable to that of a solution of the monomeric units making up the polymer matrix. For example, the "apparent" intrinsic viscosity (intrinsic microviscosity, $[\eta^\mu]$) of the various dextrans was evaluated

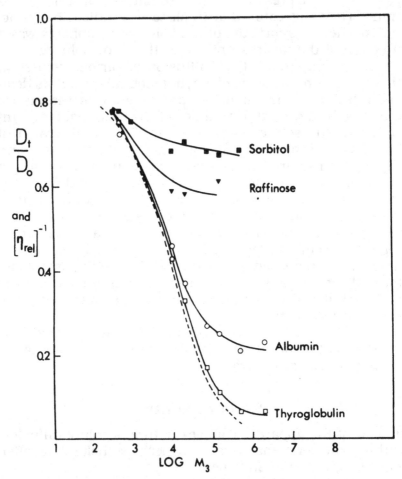

Fig. 7. Reduced diffusion coefficients of various solutes in a saccharide matrix of concentration 97.0 mg mL$^{-1}$. The molecular weight of the matrix material varied from that of sucrose to that of dextrans of $10^3 + \bar{M}_w < 2 \times 10^6$. There is an obvious dependence of solute diffusion on the molecular size of matrix material. The reduction in solute diffusion paralleled the increase in macroscopic viscosity [here plotted as $\eta_{rel}^{-1}$ (---)] in systems in which the size of solute was greater than that of matrix component. In situations in which the molecular size of diffusant was less than the size of the matrix material, its diffusion rate became independent of the size of the matrix component.

as between 3.0 and 3.8 mL g$^{-1}$ (Fig. 4), which may be compared with the intrinsic viscosity of sucrose of 2.60 mL g$^{-1}$. Obviously from Fig. 4, the value of [$\eta^\mu$] exhibits a slight dependence upon the size of diffusant. It may be simplisitic to visualize that the small diffusant only "sees" segments of the network comparable to its own size. It is interesting to note that the "scaling" treatment of Langevin and Rondelez (1978) predicted that the low molecular weight solutes would experience a frictional coefficient related to the viscosity of the solvent.

(b) GLOBULAR MACROMOLECULAR SOLUTES OF INTERMEDIATE SIZE   The reduction in transport rates of globular macromolecular diffusants of intermediate size ($10^4$ < molecular weight ≤ $10^6$ or 1.0 nm < r ≤ 10 nm) in networks displays a complex dependency. The factors controlling this dependency are (i) molecular size of diffusant and (ii) molecular size and concentration of the polymer species forming the matrix. Although both the obstacle theory of Ogston et al. (1973) and the "scaling" treatment of Langevin and Rondelez (1978) describe adequately the dependence of the retardation on the matrix concentration, neither approach can accommodate a dependency on the molecular weight of the matrix material. It is observed that when the size of diffusant is comparable to, or greater than, the size of the matrix polymer, then the macroscopic viscosity of the polymer solution describes the frictional interactions. In regimes where the size of diffusant is smaller than that of the matrix polymer, the resistance becomes effectively independent of network polymer size.

(c) LARGE PARTICLES   In the case where the size of diffusant is very large, we return to the situation of the macroscopic viscosity being an adequate representation of the frictional interactions, as predicted by the scaling treatment (Langevin and Rondelez, 1978).

(d) ASYMMETRIC MACROMOLECULAR SOLUTES   The transport behavior of asymmetric particles in matrices is even more complex. Considerations of end-on movement [reptation; De Gennes (1971)] must be undertaken. The situation at this stage of our knowledge is unclear, especially in view of our recent discoveries (Preston et al., 1980).

## Rotational Diffusion

In addition to translational diffusion, molecules also exhibit rotational movements. It is difficult to visualize that the rotational diffusion of a spherical molecule situated in a polysaccharide matrix could be impeded by obstacles. A very different relationship between rotational diffusion constant and matrix concentration is expected than in the case of translational movement.

Laurent and Öbrink have studied the rotational diffusion of serum albumin and some other proteins in concentrated dextran solutions (Laurent and Öbrink, 1972; Öbrink and Laurent, 1974). Preston et al. (1973) made a similar study of the rotation of albumin in the presence of connective tissue polysaccharides. The polymers have only a small effect on the rotation even at very high concentrations (up to 40% in the case of dextran). Furthermore, all dextrans of different molecular weights, and even sucrose, had approximately the same effect on the motion. The conclusion was that the proteins were situated in holes in the matrix where they could rotate freely, but from which they were unable to escape.

In view of the discussion above about intrinsic microviscosity in polymer solutions, we have plotted the data of Laurent and Öbrink on the rotational diffusion constant as we did with translational diffusion in Fig. 8. The plot leads to an intrinsic microviscosity of about 3–4 mL $g^{-1}$, in very close agreement with the value for translational movement of low molecular weight compounds in dextran (Fig. 4). A molecule rotating in a matrix thus is retarded by the "local viscosity," but it does not recognize obstacles or macroscopic viscosity.

## Transport of Electrolytes

A principal aspect in elucidating the transport behavior of connective tissue anionic polysaccharides is the evaluation of the effect of their high charge densities on the characteristics of their own transport, together with the transport of microions in their vicinity. The anionic polysaccharides are known to exert a significant influence on the electrochem-

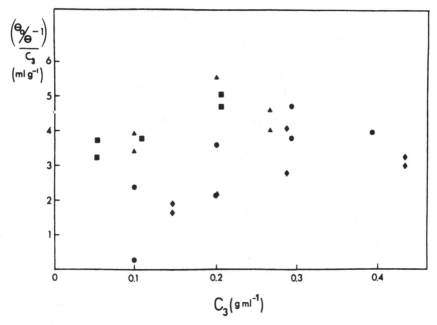

Fig. 8. The data from Laurent and Öbrink (1972) on the rotational diffusion of human serum albumin in sucrose and dextrans have been replotted to estimate the "intrinsic microviscosity" that the rotation of the protein encounters [compare Eq. (15)]. $\theta$ is the rotational diffusion constant in the presence of saccharide of concentration $c_3$, and $\theta_o$ that in solvent: sucrose ($\blacklozenge$); dextran, $\bar{M}_w \sim 10^4$ ($\blacksquare$); $4 \times 10^4$ ($\bullet$); $5 \times 10^5$ ($\blacktriangle$).

ical properties of tissues [for references, see Comper and Laurent (1978)]. Many of these properties, including transport phenomena, can be understood from the equilibrium parameters describing microion–polyanion interactions. It is worthwhile then to briefly discuss the nature of these interactions. Most of the information on transport phenomena in these systems has come from model systems. This approach has afforded a good understanding of the important features of the system. In that the majority of studies have utilized "dilute" polymer systems, it is to be noted even in these systems, when dealing with microion–polyanion interaction with polyion, that they can be classed as concentrated solutions. For example, the counterion concentration in the immediate vicinity of a polyion such as chondroitin sulfate can be as high as $4 \times 10^{20}$ equiv $L^{-1}$.

## Simple Electrolytes

The work of Manning [for a review see Manning (1979)] seems to provide a satisfactory theoretical framework for which the electrostatic interactions of microions with the anionic polysaccharides can be studied. Manning has developed expressions to describe the thermodynamic and transport properties of microions in polyelectrolyte solutions directly in terms of the structure of the polyelectrolyte. The theory is developed for a state of infinite dilution and the relationships derived are therefore considered as "limiting laws," although excellent agreement with theory has been found on studies with real solutions.

In assembling counterions onto the polyelectrolyte it can be demonstrated that a critical charge density ($\xi_{crit}$) exists on the idealized "cylindrical" polyion, above which the counterions will associate onto the surface of the cylinder so that the net charge density (inclusive of the charge of the counterions) is maintained at a critical value. The charge–density parameter $\xi$, which forms a central part of Manning's theory, is defined by:

$$\xi = e\beta/EKT \qquad (19)$$

where $\beta$ is the charge contained in a unit length of cylinder, $e$ is the charge of an electron, $E$ is the dielectric constant of the bulk solvent, $K$ is the Boltzmann constant, and $T$ is the temperature. Furthermore, the uniform charge density $\beta$ is given by:

$$\beta = Z_p e/b \qquad (20)$$

where $b = L/P$, and $L$ is the contour length of the polyion that carries $P$ charged groups of valence $Z_p$. Note that $\beta$ and therefore $\xi$ can be evaluated from the known structure of the particular polymer under study. The value of $\xi_{crit}$ for monovalent charged groups ($Z_p = 1$), as for glycosaminoglycans, is $\xi_{crit} = Z_i^{-1}$ where $Z_i$ is the valence of the counterion. Therefore, if $\xi > \xi_{crit}$ then the counterions will condense onto the polyion until $\xi = \xi_{crit}$. The condensed counterions have little or no translational mobility relative to the polyion. However, they may move freely within the electrostatic envelope of the polyion (Manning, 1979). Subsequent to the condensation of counterions onto the polyelectrolyte surface, there is a high residual field caused by the net charge of the polyion and it affects the uncondensed or "free" ions so that their behavior is modified. The interaction of the uncondensed counterions with the polyelectrolyte can be described by the Debye-Huckel mode of interaction in the limiting case. This type of interaction is known to contribute differently in a quantitative manner to various types of measurement of counterion behavior, and has lead to a misinterpretation of "binding" when its effects have not been considered. An empirical extension of the Manning treatment to finite salt concentrations has been carried out by Wells (1973b) by taking into account the interactions between mobile microions that normally occur in the absence of the polyion. The question of rates of exchange between condensed and uncondensed counterions is unresolved. It is important to note that there is no implication that condensed counterions are in condensed states for long times, or at discrete sites, or involved in specific binding. The differences of interaction of various ions with polyanions in Manning's theory are based solely on their valence. The situation may be complicated in certain systems by the occurrence of "site" binding or specific binding.

It is clear from Eq. (20) that the value of $\beta$ primarily determines the type of interaction between microion and polyion. For biological systems, their behavior will be determined by the linear charge density of the polyanion, and in particular by the dielectric constant of the system. Though there is insufficient information to assess biological systems in this regard, there are many problems to be addressed in relation to the transport of charged molecules, e.g., transfer of nutrients, metabolites, anabolic and catabolic products of cells, biosynthesis of charge polysaccharides, and organizational requirements in extracellular, intracellular, and intramembrane regions. It is noteworthy that many of these regions offer different electrochemical environments for charge interactions to occur, e.g., low dielectric constant of lipid regions, and low pH and ionic strength found in various cell granules.

In discussing the principles of microion diffusion in polyanion solutions or compartments, a qualitative understanding can be obtained within the framework of

Manning's limiting laws. It is known that condensed counterions have zero mobility with respect to the polyion [for a review, see Manning (1979)]. The kinetic, uncondensed microions will move within the electric field envelope of the polyion. The self-diffusion of both the counterion and co-ion in terms of their total concentration are predicted to be diminished relative to their values in the absence of polyion. When counterion condensation occurs the lowering of the self-diffusion coefficient of the counterion is greater than that for the co-ion. A number of studies have been performed on the self-diffusion of monovalent and divalent ions in glycosaminoglycan systems that have demonstrated general consistency with the Manning theory. The values of $b$ obtained by these measurements are compared to the theoretical values in Table 1.

In discussing the transport of microions across polysaccharide systems or membranes, the situation becomes more complex because of the presence of the Donnan distribution at each membrane–solution interface. As a result of the relatively higher concentration of counterion in the membrane, the relative differences in counterion and co-ion flux become marked at high concentration of fixed membrane charge or at low ionic strength. The unidirectional flux of the counterion tends to be higher, and that of the co-ion lower, than in the absence of the polyion. A comprehensive study of ion transport across membranes containing glycosaminoglycans, under zero and nonzero chemical potential gradients of simple electrolyte, have been made by Comper and Preston (1975). These results have been used to predict the behavior of microion and electrolyte transport in intercellular matrices of tissues of varying charge density, and are depicted in Fig. 9. For tissues in contact with physiological saline, it was predicted that only for the most densely charged tissues, such as the intervertebral disc and nasal septum, where the average fixed-charge concentration may be as high as 0.6 mol dm$^{-3}$, are marked effects on ion transport to be expected through polyion-mobile ion interaction. It is estimated that by regarding these tissues as membranes then Na$^+$ ion permeability across such tissues could be increased 2–3 times, the Cl$^-$ ion permeability retarded to 0.2–0.3 times that in free solution, while sodium chloride migration is reduced to approx-

TABLE 1
Comparison of Mean Intercharge Distances of Several Polysaccharides Estimated from Structural Models and from Counterion Intra-diffusion Data

| Polymer | Mean intercharge distance, nm | | | Reference |
|---|---|---|---|---|
| | Structure | $Na^+$-intradiffusion | $Ca^{2+}$-intradiffusion | |
| Heparin | 0.24 | 0.262 | — | Ander et al. (1978) |
| Bovine nasal septum proteoglycan subunit | 0.54 | 0.54 | 0.45 | Kitchen (1975) |
| Hyaluronate | 1.20 | 1.20 | — | Wells (1973b) |
| Chondroitin sulfate | 0.51 | 0.64 | — | Preston et al. (1972) |
| | 0.51 | 0.56 | 0.64 | Magdelenat et al. (1974) |
| Dextran sulfate | 0.23 | 0.23 | 0.23 | Kitchen (1975) |
| Carboxymethyl cellulose: degree of substitution | | | | |
| 0.78 | 0.66 | 0.59 | 0.59 | Kitchen (1975) |
| 1.05 | 0.49 | 0.49 | 0.49 | |
| 1.33 | 0.39 | 0.39 | 0.39 | |

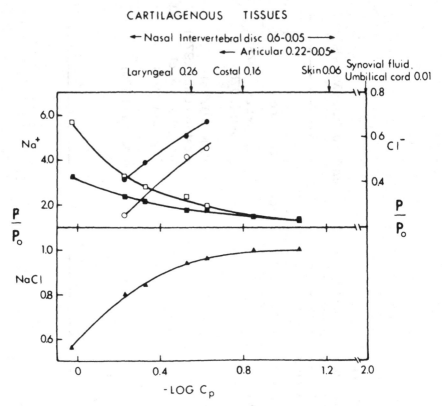

Fig. 9. Reduced permeability coefficient ($P/P_o$) for the transport of Na$^+$ [influx(□); outflux (■)], Cl$^-$ [influx (○); outflux (●)], and NaCl (▲) across membranes containing intervertebral disk proteoglycan. The charge concentration caused by the proteoglycan is $C_p$ equiv L$^{-1}$ and the membrane separates solutions of NaCl of concentrations $C_s$ and $0.5C_s$. The results have been predicted from experimental data (Comper and Preston, 1975), assuming the value of $C_s$ to be that of physiological saline. Estimates of the total fixed charge concentration of various connective tissues are included (these values are shown in brackets and have units of equivalents per liter of total tissue water).

imately 0.8 times its value in water. For articular cartilage, where the fixed-charge concentration varies from 0.05 to 0.2 mol dm$^{-3}$, values for the transport of Na$^+$ and Cl$^-$ suggest that the interaction of these ions with glycosaminoglycans of this tissue has little if any effect on their transport.

## Polyelectrolytes

There is a paucity of data concerning the transport properties of connective tissue polysaccharides. However, recent studies by Wik and Comper (1982) have been performed on the diffusion of hyaluronate at different ionic strengths. The general feature of their results is that the mutual diffusion of sodium hyaluronate can be analyzed by Eq. (3). With this particular molecule, the thermodynamic factor ($Q$) far outweighs the frictional term at high concentrations. As a result, conditions have been met where the mutual diffusion coefficient of the hyaluronate can reach values of up to $30 \times 10^{-7}$ cm$^2$ s$^{-1}$, a value that approaches the diffusion coefficient of its monomer disaccharide unit in free solution.

An interesting biological aspect of polyelectrolyte solutions is the relationship between the movement of uncondensed counterions and polyions. Normally, the movements of uncondensed counterions and of the polyelectrolyte are vastly different (an order of magnitude in most cases). The tendency for the uncondensed counterioins to move away from their opposite charge (by thermal kinetic energy) on the polyanion will create an electric potential. The existence of steady state potentials arising from concentration gradients of polyelectrolytes can be rationalized on the analysis of diffuse junctions on a long time scale ( $> 10^{-9}$ s) and transient charge distributions that do not violate the electroneutrality condition. A steady-state charge separation in this aqueous system can be viewed to be an analogous situation to piezoelectric phenomena in the solid phase. Piezoelectricity generated from collagen matrices has been well established [for a review, see Brighton et al. (1979)]. Initial studies in this direction for connective tissue polysaccharides have been made by measurement of the diffusion potentials generated by concentration gradients of chondroitin sulfate (Comper et al., 1976). The diffusion potentials measured were of considerable magnitude (up to 26 mV) for low concentration gradients (4:1 weight ratio). The potentials were larger when the diffusion of chondroitin sulfate was performed in a polymeric network generated by dextran. The network selectively generated a greater difference between counterion and polyion mobility. It was predicted that these potentials could arise under physiological conditions, and may relate to the concentration gradients of polysaccharides in tissues, and confer specific vectorial electric forces within their area.

## Formation and Transport of Matrix Structures

We have recently described a new mode of rapid macromolecular transport in concentrated polymer systems (Preston et al., 1980; Preston et al., 1984a,b; Laurent et al., 1984; Comper et al., 1984b). Although the phenomenon is not yet understood in detail, and its occurrence in biological systems has not yet been demonstrated, it is our belief that it might be of major importance for many different biological processes. In view of our limited knowledge of the reaction we will first describe our observation and then discuss tentative implications.

### Observation of Very Rapid Polymer Transport in Concentrated Ternary Systems

Several years ago Preston and coworkers observed anomalous diffusional behavior of flexible chain molecules in polysaccharide matrices (Preston and Snowden, 1973b; Kitchen, 1975). Polymers such as polyethylene glycol, polyvinyl alcohol, and polyvinylpyrrolidone (PVP) migrated with an enhanced rate compared to free diffusion. This is shown in Fig. 10 in which measurements of the apparent diffusion of PVP in increasing concentrations of dextran are shown. The measurements were performed by following the transport of PVP across a boundary where the initial PVP concentration gradient was 5 mg mL$^{-1}$. The dextran concentration was uniform throughout the system. The apparent diffusion coefficient of PVP was slightly decreased by low dextran concentrations. At higher matrix concentrations it was increased to values 10–15 times those encountered in free solution. At even higher concentrations, it decreased again. The behavior was previously interpreted in terms of an interplay between the frictional and thermodynamic factors in the diffusion equation (see above) (Laurent et al., 1979; Sundelöf, 1979). The analyses were performed with a capillary technique that precluded a careful study of the kinetics of the reaction. However, it was found that the apparent diffusion coefficient of PVP was time-dependent. The transport of low molecular weight solutes, when measured simultaneously with the PVP migration, displayed the expected diffusional behavior in that their rates were reduced in the presence of the dextran matrix (Preston et al., 1984a,b; Laurent et al., 1984).

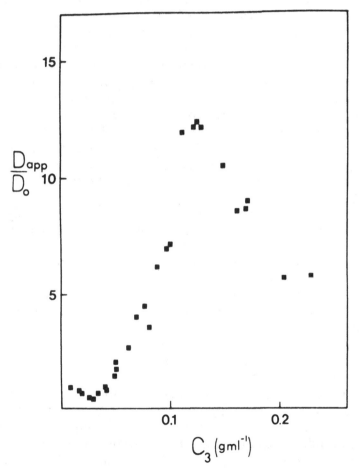

Fig. 10. The reduced apparent diffusion coefficient $(D_{app})$ of polyvinylpyrrolidone $(\bar{M}_w \sim 3.6 \times 10^5)$ (PVP) in a matrix of dextran $(\bar{M}_w \sim 10^4)$; $c_3$ = dextran concentration. $D_{app}$ was calculated assuming the PVP migration to obey normal diffusion kinetics.

We have recently made a detailed kinetic analysis of the rapid transport of PVP in the presence of dextran with the aid of diffusion cells in which horizontal boundaries are formed by a shearing mechanism (Preston et al., 1980; Preston et al., 1984b; Laurent et al., 1984). The cells were built by Professor Lars-Olof Sundelöf, Uppsala, and are described elsewhere (Sundelöf, 1984). The experiments were performed in a matrix of dextran (molecular weight $\approx 10^4$) at a concentration of 13% w/w, where a maximum in the apparent diffusion coefficient of PVP was previously observed (Fig. 10). The transport of PVP over the boundary did not follow normal diffusional kinetics, i.e., it was not linear with the square-root of time (Fig. 11). Instead the transport appeared to be linear with time, and continued until about 40% of the PVP had passed over the boundary; then it leveled off. The transport rate of sorbitol, which was measured simultaneously, followed normal diffusional kinetics and was much slower than that of PVP. This seemed to preclude the possibility of a convective process in the cell.

A transport rate linear with time suggested that the migration in some way was driven by gravity, and hence comparable experiments were performed in the analytical ultracentrifuge using ultraviolet optics to visualize the PVP. The transport rate increased with increased $g$-values, but not in a simple relationship. At 44,000$g$ the transport was only six times faster than at unit gravity.

A few significant observations emerged from the kinetic analyses: (a) the transport rate was independent of temperature if a correction was allowed for the change in viscosity of water with temperature; the migration does not therefore seem to be an activated process; (b) the transport rate of PVP was not affected by low molecular weight compounds that were introduced in the lower chamber to stabilize the boundary (e.g., 100% $D_2O$ or 1% sorbitol); (c) the transport rate was affected by gradients of high molecular weight compounds across the boundary. In most experiments the concentration difference in PVP was held at 5 mg mL$^{-1}$. The transport rate increased slightly at higher concentration gradients. It decreased at lower gradients, and at values of 0.3 mg mL$^{-1}$ PVP seemed to obey normal diffusion kinetics. As transport of PVP in the system was apparently governed by the presence of the matrix, it became apparent that one should also study whether the "diffusion" of

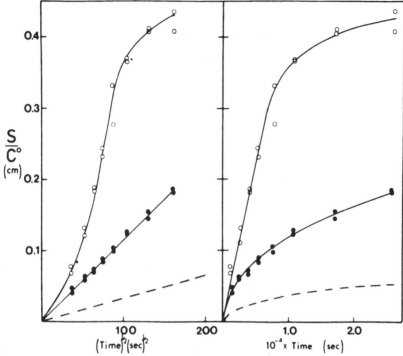

Fig. 11. The transport of polyvinylpyrrolidone (PVP) (○) and sorbitol (●) over a boundary formed by layering 135 mg mL$^{-1}$ of dextran ($\bar{M}_w \sim 10^4$) over 135 mg mL$^{-1}$ of dextran + 5 mg mL$^{-1}$ of PVP + 10 mg mL$^{-1}$ of sorbitol. S is the amount transported over the boundary per unit area and c° is the initial concentration below the boundary. Sorbitol is transported by normal diffusion kinetics, i.e., S/c° is proportional to (time)$^{1/2}$. PVP is transported more rapidly and S/c° appears to be linear with time. For comparison the normal mutual diffusion of PVP in water is included (---).

the matrix molecules behaved anomalously. Tritium-labeled dextran was therefore introduced both in the top and bottom compartments, respectively, under standard conditions, and the flux of dextran was measured in both directions over the boundary. It was found that there were considerable dextran fluxes in both directions (Fig. 12). They were much higher than would be expected from normal diffusion. The net flux of dextran over the boundary was, however, low and negative and compensated for the positive PVP flux.

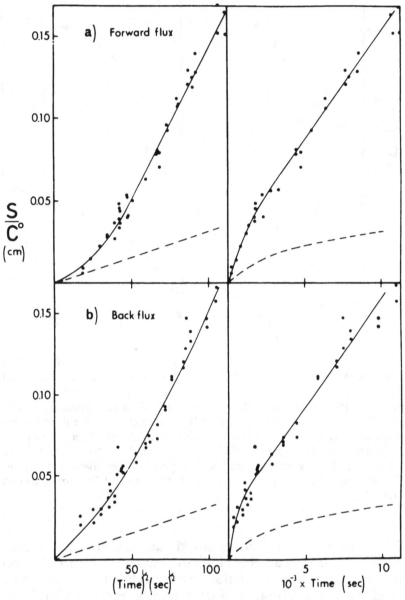

Fig. 12. The transport of dextran over the boundary described in the legend to Fig. 11. The dextran fluxes, measured by [$^3$H]-dextran, is rapid in both directions, but the net flux is small. For comparison the normal intradiffusion of dextran at $c_3 = 135$ mg mL$^{-1}$, has been included (---).

A few compounds of different molecular weights have been introduced in trace amounts in the standard system and their fluxes have been determined both along the PVP gradient and in the opposite direction. It was found that they moved with approximately the same rate in both directions, and that the transport rate increases with molecular size. Low molecular weight compounds move essentially as expected for regular diffusion (see Fig. 11) (Laurent et al., 1984).

## Visualization of the Anomalous Transport

The anomalous transport could be visualized by labeling the PVP with a blue dye (Preston et al., 1980; Comper et al., 1984). This revealed a remarkable behavior of the system. After layering a transparent dextran solution on top of the dextran solution containing 5 mg mL$^{-1}$ of blue PVP, structures were formed at the boundary which grow with constant rate. Blue "fingers" moved up in the transparent solution and transparent fingers moved down in the bottom solution. The growth of the fingers was a few millimeters per hour until they penetrated the whole system. Regular vertical columns could then be distinguished for at least 24 h before they faded away (Fig. 13). In view of the results from the transport studies one may expect that the blue fingers growing upwards contain both PVP and dextran, and the fingers growing downwards contain dextran. When the experiment was made in petri dishes with thin layers of polymer solutions and the structures were observed from above, very intricate patterns were seen transiently before the blue dye became distributed homogeneously in the solution (Fig. 14).

We have been able to visualize similar structures in a number of different ternary boundary systems. This has been accomplished both by staining the matrix polymer or the gradient-forming compound. The structures have been seen in ternary systems containing polysaccharides, nucleic acids, and proteins (Comper and Preston, 1981; Comper et al., 1984).

## A Tentative Explanation to the Phenomenon

The following might be a physical picture of the reaction. As a result of concentration fluctuations at the boundary, domains are formed with lower density than the top solution

a                                    b

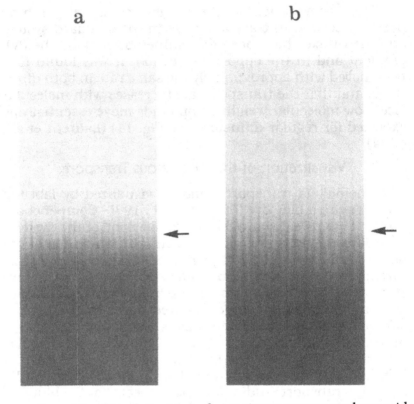

Fig. 13. The formation of structures occurs under a wide variety of experimental conditions. For example, structures may form in the presence of dextran gradients. These structures are much more ordered and coherent than structures formed in the absence of a dextran concentration gradient. This figure demonstrates the formation of structures at the boundary between 112 mg mL$^{-1}$ dextran (top) and 5 mg mL$^{-1}$ blue PVP, 135 mg mL$^{-1}$ dextran (bottom). The boundary was formed in the center of a Tiselius cell (length 6.0 cm) and photographed (a) 3 h and (b) 28 h after layering. The structures remained in the cell for at least 5 d. The initial position of boundary is shown by arrows. Magnification × 3.75.

or higher density than the bottom solution (Laurent et al., 1984; Comper et al., 1984). These density inhomogeneities start to move in the gravitational field; this is schematically shown in Fig. 15. The high molecular weight material with low diffusion coefficient remains in the moving structure, while the low molecular weight material oscil-

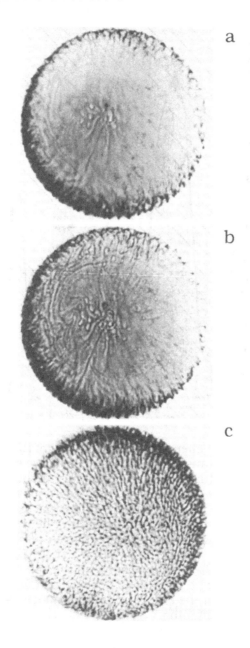

Fig. 14. The formation of structures in a petri dish, seen from above. A boundary between 135 mg mL$^{-1}$ dextran (top) and 5 mg mL$^{-1}$ blue PVP, 135 mg mL$^{-1}$ dextran and 10 mg mL$^{-1}$ sorbitol (bottom) was formed by layering 1 mL of blue solution under 1 mL of clear dextran solution with a needle. The photographs were taken (a) 4 min, (b) 12 min, and (c) 20 min after the layering. Magnification × 1.7.

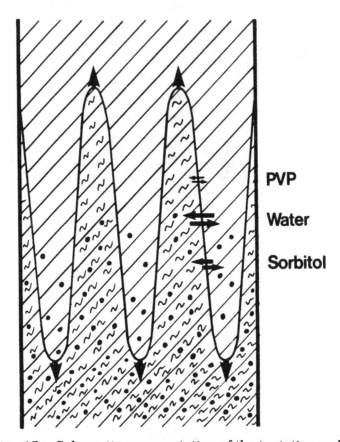

**PVP**

**Water**

**Sorbitol**

Fig. 15. Schematic representation of the tentative explanation of rapid transport involving dissipative structures. Because of density fluctuations in the boundary, structures start to move vertically in both directions. The upward moving structures contain PVP (shown as ~), but the diffusional exchange of PVP with the downward moving structure is slow hence the net PVP transport in the upward vertical direction is rapid. Conversely, with low molecular weight compounds, such as water or sorbitol (shown as ●), the exchange between the oppositely moving structures is rapid and thus their net vertical migration is slow compared to that of the PVP. The dextran matrix is here represented by the cross hatching.

lates between the ascending and the descending structures and thus its net transport in the vertical direction is low.

The phenomenon can also be explained in terms of recent developments in the theory of nonlinear non-equilibrium thermodynamics (Nicolis and Prigogine, 1977). These authors have afforded an analysis of the general conditions required for the formation of ordered, coherent structures in an open system. The existence of two basic features is necessary for this to occur. Firstly, constraints must be placed upon the system that drive it further from an equilibrium state. Secondly, nonlinear mechanisms must be present that act between the various elements of the system. When these conditions are met, certain types of fluctuations can be amplified and drive the system to a new regime different from the initial reference state. This regime is characterized by the appearance of organizations ("dissipative" structures) whose probabilities of occurrence at equilibrium would be negligible. Such structures have been described in other fields of chemistry (Tyson, 1976), but have, to our knowledge, not been demonstrated in polymer solutions. Although the movements of the structures in the presently described system are apparently driven by gravity, there are no reasons that other forces such as hydrodynamic flow could not cause similar structures to be formed.

## Biological Implications

Our observations might have a general importance in the formation of biological structures and in biological transport. In a system that can continuously generate gradients by synthesis of material, one may expect the formation of permanent structures stablized by the steady-state conditions that prevail. The scale in time and space could vary considerably from system to system. For example, the structures we have observed in polymer systems have a close resemblance to the cytoplasmic streaming seen in cells, although the latter has microscopic dimensions. Furthermore, it may seem improbable to have moving structures in the rigid gel-like matrix formed outside the cell by the interpenetrating collagen and polysaccharide networks (e.g., in cartilage). However, on a longer time-scale it is not unreasonable to believe that the matrix can move and form long-lived structures because of the very slow diffusion rate of the matrix components. Outside load on the tissue could have a direct influence on the formation of the structures. It is known that the collagen distribution in bone is determined by the load.

Transport in the nerve axon has been studied in detail [for reviews see Elam (1979) and Lasek (1980)] but it has not been explained how flow can occur in both directions, and how particulate material is transported more rapidly than soluble proteins and low molecular weight material. It is interesting that these features are exactly the type of transport features that we have demonstrated in our polymer system (Comper et al., 1983). Recent observations on the transport of high molecular weight material in certain membranes [Cumming et al. (1979); Preston and King, unpublished observations] have indicated a surprisingly high transport rate compared to free diffusion of the compounds. It will be interesting to investigate whether transport by dissipative structures can occur also in regular membranes.

## Acknowledgments

We acknowledge the large amount of experimental work carried out by Robert Kitchen, Wayne Connors, Gregory Checkley, and Geoffrey Wilson.

# Chapter 7

# Glycosaminoglycan Chains in the Biological State

## Summary of Conclusions from Physical Measurement

## D. A. Rees

National Institute for Medical Research, Mill Hill,
London, UK

### Introduction—Molecular Species

The molecular components of the extracellular matrix, which are the main topics of discussion in this volume, are hyaluronate (a linear all-carbohydrate polymer of very high molecular weight) and the family of proteoglycans, which are similar to each other in general structure, but differ in their carbohydrate side chains, their molecular weights, their capacities for interaction through polypeptide moities, and probably in other respects as well. There is now good evidence that proteoglycans may have a domain structure: for example, in chondroitin 4-sulfate proteoglycans from cartilage we distinguish (pp. 4–8) a globular headgroup (which binds hyaluronate), N-linked keratan sulfate chains in association with a distinct region of the polypeptide, and O-linked chondroitin sulfate chains attached to a region of the polypeptide that is further from the headgroup. The overall size and osmomechanical properties (Wells, 1973a), which relate to some of the biological functions (see below), are determined chiefly by the carbohydrate chains, whereas the proximity of these chains to each other in space is constrained by the polypeptide backbone, and both carbohydrate and protein moities can engage in specific noncovalent interactions (pp. 11ff, 53ff, 95ff).

## Molecular Shapes

Perhaps the most dramatic of recent advances has been in the determination by X-ray fiber diffraction analysis of glycosaminoglycan reference structures in great number and considerable detail, showing many specificities of molecular interaction. Although the models have not led immediately to direct revelations of structure–function relationships, there is every prospect (as argued below) that biological function will be much better understood when the structural detail is considered in the context of biological and other physical studies over the next few years. These advances have been made possible by important improvements in methods for preparation of polysaccharide materials in crystalline form as a starting point for analysis (Atkins and Mackie, 1972), new understanding of types of structural disorder and their influence on diffraction behavior, and systematized molecular model building and refinement by computer methods, especially using linked-atom calculations within noncontroversial stereochemical constraints (pp. 44–45). Models can then be derived that are much more precise than previously achievable and show intra- and interchain hydrogen-bonding patterns, cation positions and their coordination geometry, and even the position and hydrogen bonding of associated water molecules in favorable cases.

The learning period for method development over the first part of the decade made it necessary to revise rather radically some of the early trial structures; for example, we now know that a hyaluronate double helix does indeed exist (Sheehan et al., 1977; Arnott et al., 1983), but does not correspond to the diffraction diagram for which it was first proposed (Dea et al., 1973; Atkins and Sheehan, 1973) and that one of the dermatan sulfate allomorphs may have the right-handed chain sense (p. 60) rather than, as first proposed, all forms having the left-handed sense as is common for other glycosaminoglycans. This experience has shown the importance of carrying structure determinations through to the final stages of rigorous testing and refinement—only then can we draw firm and detailed conclusions about conformations and interactions; indeed, first impressions from trial models can mislead. However, we can now finally appreciate the great scope that these chains have for adopting orderly and well-defined shapes and for patterns of intramolecular hydrogen bonding and interchain associations to stabilize them (pp. 53–65).

Almost all glycosaminoglycan chains have a truly striking ability to exist in a series of distinct, alternative conformations in the condensed phase, most of which (p. 42) are extended hollow helices for which the "degree of winding" falls within quite a narrow range (rotation per disaccharide residue is usually between $-90°$ and $-225°$). The greatest versatility is shown by hyaluronate, for which no less than six forms have so far been characterized, and the least by keratan sulfate, for which only one form is known so far. The chondroitin sulfates and dermatan sulfates are each known in three forms. A major determinant in triggering the conversion of one conformation to another in the condensed or fiber state can evidently be the salt form; $Ca^{2+}$ is an especially strong discriminant (e.g., in converting hyaluronate from $4_3$ to $3_2$ and chondroitin 6-sulfate from $8_3$ to $3_2$; p. 62), but changes from one univalent counterion to another can also be important (p. 56). When different glycosaminoglycans exist in a similar conformation (for example, hyaluronate, chondroitin 4- and 6-sulfates, and dermatan sulfates as $3_2$ helices), the structures appear to be closely isomorphous; however, corresponding structures may require different counterions for stability. All this points to subtle and specific interactions in which variations in backbone structure and ionic environment mutually interact to establish a wide range of quantized conformational states. The most extensive studies have been made with free glycosaminoglycan chains, but in those intact proteoglycans that have been studied, the polypeptide backbone does not impose major additional constraints on polysaccharide shapes and interactions (p. 65).

Such satisfactory advances from the diffraction analysis of glycosaminoglycans of the hyaluronate-chondroitin-dermatan-keratan family have not been matched in the heparin-heparan series. Clearly these must be an important focus for definitive structural studies in the future.

When all the X-ray diffraction results are compared with those from hydrodynamic and spectroscopic studies in solution, it becomes apparent that the exquisitely ordered forms of glycosaminoglycans in the condensed phase can be substantially modified in aqueous environments. For example, detailed studies on the solution properties of hyaluronate (p. 23) have shown it to be an extended random coil, but stiffened significantly, suggesting that some features of the inter-residue relationships might be retained from the solid state even if the chain conformation is not locked rigidly in an ordered state. This model is fully supported by NMR studies on the polymer and on oligomers (p. 88ff) which further point to a hydrogen bonding pattern that is consistent with these expectations. For the L-iduronate-containing glycosaminoglycans—dermatan sulfate, and heparin and heparan sulfates—NMR shows that the conformation can be modified still further in solution by conversion of the uronate residue from the $^4C_1$ chair that exists in the known crystalline fibers to a ring form that has interproton coupling constants corresponding to those expected for $^1C_4$ (p. 78). When this occurs we do not understand how the ring conformation relates to the relative susceptibility of this residue to periodate oxidation. Taken alone, the periodate susceptibility might suggest that the $^4C_1$ conformer survives in solution, but the conflict between this evidence and NMR argues strongly against such a straightforward interpretation (p. 84). NMR shows that $Ca^{2+}$ can cause a conformational perturbation of heparin in solution (p. 89), recalling (see above) the diffraction evidence for a similar effect on other glycosaminoglycans in the condensed phase. Even here, the spectral line widths (compare p. 92) show that the chains remain mobile and therefore cannot assume the degree of rigidity characteristic of other polysaccharides that retain their solid-state conformation into solution (compare Ablett et al., 1982).

The most convincing evidence for conformational ordering in solution that might directly parallel the condensed phase is in the growing number of examples of the self-association of glycosaminoglycan chains. Three instances are now characterized, all of which can arrest at the level of dimerization. This provides good evidence for a specific rather than nonspecific interaction since the latter would proceed through to large aggregates, and is strongly reminiscent of the dimerizations that can take various geometrical forms in other polysaccharide systems (Morris et al., 1978, 1980a, 1982). The two glycosaminoglycans for which long-lived associations of this type have been clearly demonstrated by chromatography and light scattering (pp. 97ff, 106) both have a rather special composition in that they are dermatan sulfate and heparan sulfate fractions in which the D-glucuronate- and L-iduronate-containing disaccharide residues are present in mixed compositions. Other sequences including chondroitin sulfate and keratan sulfate show the behavior weakly or not at all. The association is retained at the level of proteoglycan for at least one example (dermatan sulfate proteoglycan; p. 115). The evidence that binding requires a minimum oligosaccharide chain length (five disaccharide residues) shows that the mechanism is cooperative and must involve a specific stereochemical fit of a type that should be amenable to characterization by X-ray diffraction. Unfortunately, however, the polymers that show this behavior in solution have not yet yielded to diffraction analysis and little can therefore be said at this stage about the geometry of the association.

The third example of cooperative self-association between glycosaminoglycans is for hyaluronate, but this differs from the other two systems in that the interaction is fleeting rather than long-lived. In this case the evidence is from the competitive inhibition of network formation as monitored by dynamic moduli when oligosaccharides are added (Welsh et al., 1980). Since these oligosaccharides must exceed a critical chain length to be effective and the inhibition itself cannot be accounted for by unlimited ag-

gregation or random coil interpenetration, the likelihood is
that again we have a cooperative, stereochemical fit and
that this could be modeled by a known X-ray structure. Of
the known forms, it is unlikely that the twofold ribbon is involved because network properties do not correlate with the
very low pH that (p. 56) induces this conformation in the
solid state. However, the double helix remains as likely a
condidate now as a decade ago (Dea et al., 1973), especially
since its existence in condensed films (Sheehan et al.,
1977; p. 56) parallels network development (Balazs, 1966)
in being favor ed by slight but not strong acidification. Although its formation in films shows a cation specificity not
seen in solution, recent refinement of the X-ray structure
suggests that the role of the cation is between double helix
structures rather than within them (Arnott et al., 1983)
and therefore might not be necessary for network development. Other candidates are the fourfold and threefold hollow helices (p. 42) possibly "super-hydrogen-bonded" (p. 90)
and nested side by side. Further arbitration between these
possibilities might be possible after investigation of cation
preferences for network formation in comparison with cation effects in the solid state. This evidence that
glycosaminoglycans require interchain association to stabilize ordered conformations in an aqueous environment is
fully consistent with the behavior of other polysaccharide
systems that, when adopting conformational order, typically do so in two-chain or three-chain assemblies. Apart
from the three examples we have noted, however, it is possible that glycosaminoglycans will be found to adopt ordered shapes in nature only when in tissues and tissue
compartments having low water activity, and/or in interactions yet to be characterized in detail with the proteins to
which they bind in the extracellular matrix (see below).

## Molecular Assemblies

The glycosaminoglycan self-associations may be assumed
to play a part with other interactions in holding together
the total composite structure of the extracellular matrix. No
doubt carbohydrate–protein and protein–protein interac-

tions, as well as topological interpenetrations of the large extended macromolecules, are also important. Among carbohydrate–protein interactions in connective tissues, the best characterized example is the binding of chondroitin 4-sulfate proteoglycan to hyaluronate. This also has a clear biological function in serving to hold the proteoglycan within the collagen framework in cartilage to create a swelling pressure through the osmotic activity of the chondroitin sulfate chains (Wells, 1973a), and so maintain resilience of the tissue. It has been possible to characterize this binding thoroughly because it is readily reconstructed in vitro (p. 12). Indeed its mechanisms of biological assembly have been studied in cell culture systems (p. 13) to show the striking and fascinating result that the associations are formed extracellularly *after* secretion. It is important to note, however, that not all proteoglycans show this headgroup-binding to hyaluronate (pp. 18–19), and therefore that if they are fixed in the tissue this must occur by other mechanisms. Other interactions are beginning to be recognized that could be important in basement membrane as well as cartilage and other connective tissues, but are not covered in this volume, e.g., glycosaminoglycan–fibronectin binding (Yamada et al., 1982). There is also evidence that some proteoglycans can self-associate through their protein cores (p. 38). If, as is likely, these represent assembly steps that occur in the extracellular space, we must presume that the regulatory mechanisms exist in the structures of the interacting species themselves rather than being through other controlling agents.

Related to these questions is the problem of how ions and molecules, including large molecules, move about in the matrix during and after formation. Some of the main influences on rotational and translational diffusion, and also on the activity of small counterions, have been understood for many years—for example, the importance of steric interactions between diffusant and matrix and of the linear charge density of the polymer backbone have been so well understood that some good quantitative predictions of behavior have been possible for systems in vitro (p. 136). However, unexplained effects do still remain, such as the dependence of macromolecular diffusion on the molecular weight of the matrix species (p. 138). A new development of great potential importance (p. 152) is the discovery of a mechanism by which one polymer can accelerate the transport of another in a manner that increases with the molecular weight of the diffusing species and has been demonstrated for many ternary systems including polysaccharides, nucleic acids, and proteins. This probably arises from local density differences caused by concentration fluctuations. One of the many fascinating challenges for the future will be to investigate such dynamic aspects of macromolecular interactions in parallel with structural aspects of the molecular assembly of native tissues, so that the relationships between the extracellular matrix and cellular function can be fully understood.

# References

Ablett, S., Clark, A. H., and Rees, D. A. (1982). *Macromolecules,* **15,** 597–602.

Aharoni, S. M. (1978). *J. Macromol. Sci. Phys. B,* **15,** 347–370.

Ander, P., Gangi, G., and Kowblansky, A. (1978). *Macromolecules,* **11,** 904–908.

Angyal, S. J. (1969). *Angew. Chem. Int. Ed.,* **8,** 157–166.

Arnott, S., Guss, J. M., and Hukins, D. W. L. (1973a). *Biochem. Biophys. Res. Commun.,* **54,** 1377–1383.

Arnott, S., Guss, J. M., Hukins, D. W. L., and Mathews, M. B. (1973b). *Science,* **180,** 743–745.

Arnott, S., Chandrasekaran, R., and Marttila, C. (1974a). *Biochem. J.,* **141,** 537–543.

Arnott, S., Chandrasekaran, R., Hukins, D. W. L., Smith, P. J. C., and Watts, L. (1974b). *J. Mol. Biol.,* **88,** 523–533.

Arnott, S., Guss, J. M., Hukins, D. W. L., Dea, I. C. M., and Rees, D. A. (1974c). *J. Mol. Biol.* **88,** 175–184.

Arnott, S., Chandrasekaran, R., and Leslie, A. G. W. (1976). *J. Mol. Biol.,* **106,** 735–748.

Arnott, S., Chandrasekaran, R., Bond, P. J., Birdsall, D. L., Leslie, A. G. W., and Puigjaner, L. C. (1981). *Seventh Aharon Katzir-Katchalsky Conference on Structural Aspects of Recognition and Assembly in Biological Macromolecules,* Nof Ginossar, Israel, **2,** 487–500.

Arnott, S., Mitra, A. K., and Raghunathan, S. (1983). *J. Mol. Biol.,* **169,** 861–872.

Atkins, E. D. T., and Isaac, D. H. (1973). *J. Mol. Biol.,* **80,** 773–779.

Atkins, E. D. T., and Mackie, W. (1972). *Biopolymers,* **11,** 1685–1691.

Atkins, E. D. T., and Sheehan, J. (1972). *Nature New Biol.*, **235**, 235–256.

Atkins, E. D. T., and Sheehan, J. K. (1973). *Science*, **179**, 562–564.

Atkins, E. D. T., Gaussen, R., Issac, D. H., Nandanwar, V., and Sheehan, J. K. (1972). *J. Polymer Sci.*, **B10**, 863–876.

Atkins, E. D. T., Issac, D. H., Nieduszynski, I. A., Phelps, C. F., and Sheehan, J. K. (1974a). *Polymer J.*, **15**, 263–271.

Atkins, E. D. T., Hardingham, T. E., Isaac, D. H., and Muir, H. (1974b). *Biochem. J.*, **141**, 919–921.

Atkins, E. D. T., Meader, D., and Scott, J. E. (1980). *Int. J. Biol. Macromol.*, **2**, 318–319.

Baker, J. R., and Caterson, B. (1979). *J. Biol. Chem.*, **254**, 2387–2393.

Balazs, E. A. (1966). *Fed. Proc.*, **25**, 1817–1822.

Balazs, E. A., McKinnon, A. A., Morris, E. R., Rees, D. A., and Welsh, E. J. (1977). *J.C.S. Chem. Commun.*, 44–45.

Barker, S. A., and Young, N. M. (1966). *Carbohydr. Res.*, **2**, 366–370.

Barrett, T. W. (1978). *Biopolymers*, **17**, 1567–1579.

Barrett, T. W., and Harrington, R. E. (1979). *Biopolymers*, **16**, 2167–2188.

Bearman, R. J. (1961), *J. Phys. Chem.*, **65**, 1961–1968.

Blix, G., and Snellman, O. (1945). *Ark. Kemi Mineral Geol.*, 19a, No. 32.

Bociek, S. M., Darke, A. H., Welti, D., and Rees, D. A. (1980). *Eur. J. Biochem.*, **109**, 447–456.

Bode, H.-J. (1979). *Z. Naturforsch.*, **34c**, 512–528.

Boyd, J., Williamson, F. B., and Gettins, P. (1980). *J. Mol. Biol.*, **137**, 175–190.

Brehm, G. A., and Bloomfield, V. A. (1975). *Macromolecules*, **8**, 663–665.

Brighton, C. T., Black, J., and Pollack, S. R. (1979). *Electrical Properties of Bone and Cartilage*, Grune & Stratton, New York

Cabassi, F., Casu, B., and Perlin, A. S. (1978). *Carbohydr. Res.*, **63**, 1–11.

Cael, J. J., Winter, W. T., and Arnott, S. (1978). *J. Mol. Biol.*, **125**, 21–42.

Campbell, I. D., Dobson, C. M., Williams, R. J. P., and Xavia, A. V. (1973). *J. Magn. Reson.*, **11**, 172–181.

Casassa, E. F., and Berry, G. C. (1966). *J. Polym. Sci A-2*, **4**, 881–897.

Casu, B., Reggiani, M., Gallo, G., and Vigevani, A. (1970). *Carbohydr. Res.*, **12**, 157–170.

Casu, B., Gatti, G., Cyr, N., and Perlin, A. S. (1975). *Carbohydr. Res.*, **41**, C6–C8.

Casu, B., Cifonell, A. J., Perlin, A. S., and Scovenna, G. (1978). *Carbohydr. Res.*, **63**, 13–27.

Casu, B., Gatti, G., Gennaro, U., Perlin, A. S., and Vincendon, M. (1984). In preparation.

Chou, C. H., Thomas, G. J. Jr., Arnott, S., and Smith, P. J. C. (1977). *Nucleic Acids Res.*, **4**, 2407–2419.

Christner, J. E., Brown, M. L., and Dziewiatkowski, D. D. (1977). *Biochem. J.*, **167**, 711–716.

Christner, J. E., Brown, M. L., and Dziewiatkowski, D. D. (1979). *J. Biol. Chem.*, **254**, 4624–4630.

Cleland, R. L. (1970a). In *Chemistry and Molecular Biology of the Intercellular Matrix* (Balazs, E. A., ed), pp. 733–742, Academic Press, London.

Cleland, R. L. (1970b). *Biopolymers*, **9**, 811–824.

Cleland, R. L. (1977). *Arch. Biochem. Biophys.*, **180**, 57–68.

Cleland, R. L., and Wang, J. L. (1970). *Biopolymers*, **9**, 799–810.

Comper, W. D., and Laurent, T. C. (1978). *Physiol. Rev.*, **58**, 255–315.

Comper, W. D., and Preston, B. N. (1975). *J. Colloid Interface Sci.*, **53**, 379–390.

Comper, W. D., Lisberg, W., and Veis, A. (1976). *J. Colloid Interface Sci.*, **57**, 345–352.

Comper, W. D., and Preston, B. N. (1981). *Biochem. Int.*, **3**, 557–564.

Comper, W. D., Preston, B. N., and Austin, L. (1983). *Neurochem. Res.*, **8**, 943–955.

Comper, W. D., Preston, B. N., Laurent, T. C., Checkley, G. J., and Murphy, W. H. (1984). *J. Phys. Chem.*, in press.

Comper, W. D., Van Damme, M.-P., and Preston, B. N. (1982). *J. Chem. Soc., Faraday Trans. I*, **78**, 3369–3378.

Cöster, L., and Fransson, L.-Å. (1981). *Biochem. J.*, **193**, 143–153.

Cöster, L., Malmström, A., Sjöberg, I., and Fransson, L.-Å. (1975), *Biochem. J.*, **145**, 379–389.

Cöster, L., Franssön, L.-Å., Sheehan, J., Nieduszynski, I. A., and Phelps, C. F. (1981). *Biochem. J.*, **197**, 483–490.

Crick, F. H. C., and Watson, J. D. (1954). *Proc. Roy. Soc. A*, **223**, 80–96.

Cumming, G. J., Handley, C. J., and Preston, B. N. (1979). *Biochem. J.*, **181**, 257–266.

Darke, A., Finer, E. G., Moorhouse, R., and Rees, D. A. (1975). *J. Mol. Biol.*, **99**, 477–486.

Dea, I. C. M., Moorhouse, R., Rees, D. A., Arnott, S., Guss, J. M., and Balazs, E. A. (1972). *Scand. J. Clin. Lab. Invest.*, **29** (Suppl. 123), 10.

Dea, I. C. M., Moorhouse, R., Rees, D. A., Arnott, S., Guss, J. M., and Balazs, E. A. (1973). *Science*, **179**, 560–562.

De Bruyn, A., and Anteunis, M. (1976). *Org. Magn. Reson.*, **8**, 228.

De Gennes, P. G. (1971). *J. Chem. Phys.*, **55**, 572–579.

De Gennes, P. G. (1979). *Nature*, **282**, 367–370.

De Luca, S., Heinegård, D., Hascall, V. C., Kimura, J. H., and Caplan, A. I. (1977). *J. Biol. Chem.*, **252**, 6600–6608.

De Luca, S., Lohmander, L. S., Nilsson, B., Hascall, V. C., and Caplan, A. I. (1980). *J. Biol. Chem.*, **255**, 6077–6083.

Des Cloizeaux, J. (1975). *J. de Phys.*, **36**, 281–291.

Di Ferrante, N., Donnelly, P. V., and Berglund, R. K. (1971). *Biochem. J.*, **124**, 549–553.

Donners, W. A. B. (1977). *Colloid Polym. Sci.*, **225**, 27–31.

Edelson, E., and Fuoss, R. M. (1950). *J. Amer. Chem. Soc.*, **72**, 306–310.

Ehrlich, J., and Stivala, S. S. (1974). *Polymer*, **15**, 204–210.

Einstein, A. (1905). *Ann. Phys. Lpz.*, **17**, 549–560.

Einstein, A. (1908). *Z. Electrochem.*, **14**, 235–239.

Elam, J. S. (1979). In *Complex Carbohydrates of Nervous Tissue* (Margolis, R. V., and Margolis, R. K., eds), Plenum Press, New York.

Elworthy, P. H., Florence, A. T., and Rahman, A. (1972). *J. Phys. Chem.*, **76**, 1763–1767.

Fellini, S. A., Kimura, J. H., and Hascall, V. C. (1981). *J. Biol. Chem.*, **256**, 7883–7889.

Fessler, J. H., and Fessler, L. I. (1966). *Proc. Natl. Acad. Sci. USA*, **56**, 141–147.

Flory, P. J. (1953). *Principles of Polymer Chemistry*, Cornell University Press, New York.

Flory, P. J., and Fox, T. G. (1951). *J. Amer. Chem. Soc.*, **73**, 1904–1908.

Fransson, L.-Å. (1974). *Carbohydr. Res.*, **36**, 339–348.

Fransson, L.-Å. (1976). *Biochim. Biophys. Acta*, **437**, 106–115.

Fransson, L.-Å. (1978). *Carbohydr. Res.*, **62**, 235–244.

Fransson, L.-Å, and Cöster, L. (1979). *Biochim. Biophys. Acta*, **583**, 132–144.

Fransson, L.-Å., Cöster, L., Malmström, A., and Sjöberg, I. (1974). *Biochem. J.*, **143**, 369–378.

Fransson, L.-Å, Huckerby, T. N., and Nieduszynski, I. A. (1978). *Biochem. J.*, **175**, 299–309.

Fransson, L.-Å., Nieduszynski, I. A., Phelps, C. F., and Sheehan, J. K. (1979). *Biochim. Biophys. Acta*, **586**, 179–188.

Fransson, L.-Å., Nieduszynski, I. A., and Sheehan, J. K. (1980a). *Biochim. Biophys. Acta*, **630**, 287–300.

Fransson, L.-Å., Sjöberg, I., and Havsmark, B. (1980b). *Eur. J. Biochem.*, **106**, 59–69.

Gardell, S., Baker, J. R., Caterson, B., Heinegård, D., and Rodén, L. (1980). *Biochem. Biophys. Res. Commun.*, **95**, 1823–1831.

Gatti, G. (1978). In *NMR Spectroscopy in Molecular Biology* (B. Pullman, Ed.), p. 423, Reidel, Dordrech, Holland.

Gatti, G. Casu, B., and Perlin, A. S. (1978). *Biochem. Biophys. Res. Commun., 85*, 14–20.

Gatti, G., Casu, B., Torri, G., and Vercellotti, J. R. (1979a). *Carbohydr. Res., 68*, C3–C7.

Gatti, G., Casu, B., Hamer, G. K., and Perlin, A. S. (1979b). *Macromolecules, 12*, 1001–1007.

Gibbs, D. A., Merrill, E. W., Smith, K. A., and Balazs, E. A. (1968), *Biopolymers, 6*, 771–791.

Guss, J. M., Hukins, D. W. L., Smith, P. J. C., Winter, W. T., Arnott, S., Moorhouse, R., and Rees, D. A. (1975). *J. Mol. Biol., 95*, 359–384.

Hadler, N. M. (1980). *J. Biol. Chem., 255*, 3532–3535.

Hallett, F. R., and Gray, A. L. (1974). *Biochim. Biophys. Acta, 343*, 648–655.

Hamer, G. K., and Perlin, A. S. (1976). *Carbohydr. Res., 49*, 37–48.

Hamer, G. K., Balza, F., Cyr, N., and Perlin, A. S. (1978). *Canad. J. Chem., 56*, 3109–3116.

Hammerman, D., and Schubert, M. (1962). *Amer. J. Med., 33*, 555–590.

Hardingham, T. E. (1979). *Biochem. J., 177*, 237–247.

Hardingham, T. E., and Muir, H. (1972a). *Biochem. J., 126*, 791–803.

Hardingham, T. E., and Muir, H. (1972b). *Biochim. Biophys. Acta, 279*, 401–405.

Hardingham, T. E., and Muir, H. (1973a). *Biochem. Soc. Trans., 1*, 283–284.

Hardingham, T. E., and Muir, H. (1973b). *Biochem. J., 135*, 905–908.

Hardingham, T. E., and Muir, H. (1974). *Biochem. J., 139*, 565–581.

Hardingham, T. E., Ewins, R. J. F., and Muir, H. (1976). *Biochem. J., 157*, 127–143.

Hardingham, T. E., Ewins, R. J. F., Dunham, D. G., and Muir, H. (1981). *Seminars in Arthritis and Rheumatism XI*, No. 1, Suppl. 1, 28–29.

Harper, G. S., Comper, W. D., and Preston, B. N. (1982). *Connective Tissue Res.*, **9**, 209.

Hascall, V. C. (1977). *J. Supramol. Struct.*, **7**, 101–120.

Hascall, V. C., and Heinegård, D. (1974). *J. Biol. Chem.*, **249**, 4242–4249.

Hascall, V. C., and Sajdera, S. W. (1969). *J. Biol. Chem.*, **244**, 2384–2396.

Hascall, V. C., and Sajdera, S. W. (1970). *J. Biol. Chem.*, **254**, 4920–4930.

Hawkins, E. Y., Foweraker, A. R., and Jennings, B. R. (1978). *Polymer*, **19**, 1233–1236.

Hearst, J. E., and Stockmayer, W. H. (1964). *J. Chem. Phys.*, **37**, 1424–1433.

Heatley, F., Scott, J. E., and Casu, B. (1979). *Carbohydr. Res.*, **72**, 13–23.

Heatley, F., Scott, J. E., Jeanloz, R. W., and Walker-Nasir, E. (1982). *Carbohydr. Res.*, **99**, 1–11.

Heinegård, D. (1977). *J. Biol. Chem.*, **252**, 1980–1986.

Heinegård, D., and Hascall, V. C. (1974). *J. Biol. Chem.*, **249**, 4250–4256.

Höök, M., Lindahl, U., Bäckström, G., Malmström, A., and Fransson, L.-Å. (1974). *J. Biol, Chem.*, **249**, 3908–3915.

Houwink, R. (1940). *J. Prakt. Chem.*, **157**, 15–22.

Huckerby, T. N., and Nieduszynsky, I. A. (1982). *Carbohydr. Res.*, **103**, 141–145.

Isaac, D. H., and Atkins, E. D. T. (1973). *Nature New Biol.* **244**, 252–253.

Kimata, K., Hascall, V. C., and Kimura, J. H. (1982). *J. Biol. Chem.*, **257**, 3827–3832.

Kimura, J. H., Osdoby, P., Caplan, A. I., and Hascall, V. C. (1978). *J. Biol. Chem.*, **253**, 4721–4729.

Kimura, J. H., Hardingham, T. E., Hascall, V. C., and Solursh, M. (1979). *J. Biol. Chem.*, **254**, 2600–2609.

Kimura, J. H., Hardingham, T. E., and Hascall, V. C. (1980). *J. Biol. Chem.*, **255**, 7134–7143.

Kitchen, R. G. (1975). Thesis, Monash University, Clayton, Victoria, Australia.

Kitchen, R. G., and Cleland, R. L. (1978). *Biopolymers*, **17**, 759–783.

Kitchen, R. G., Preston, B. N., and Wells, J. D. (1976). *J. Polym. Sci. Symp.*, **55**, 39–49.

Koch, H. J., and Perlin, A. S. (1970). *Carbohydr. Res.*, **15**, 403–410.

Komiyama, J., and Fuoss, R. M. (1972). *Proc. Natl. Acad. Sci. USA*, **69**, 829–833.

Kratky, O. (1966). *Pure Appl. Chem.*, **12**, 483–523.

Kurata, M., and Stockmayer, W. H. (1963). *Forschr. Hochpolym. Forsch.*, **3**, 196–312.

Langevin, D., and Rondelez, F. (1978). *Polymer*, **19**, 875–882.

Langridge, R., and Rich, A. (1963). *Nature*, **198**, 725–728.

Lasek, R. J. (1980). *Trends in Neurosciences*, **3**, 87–91.

Laurent, T. C. (1966). *Fed. Proc.*, **25**, 1128–1134.

Laurent, T. C., and Öbrink, B. (1972). *Eur. J. Biochem.*, **28**, 94–101.

Laurent, T. C., and Persson, H. (1963). *Biochim. Biophys. Acta*, **78**, 360–366.

Laurent, T. C., and Persson, H. (1964). *Biochim. Biophys. Acta*, **83**, 141–147.

Laurent, T. C., Ryan, M., and Petruszkiewiez, A. (1960). *Biochim. Biophys. Acta*, **42**, 476–485.

Laurent, T. C., Björk, I., Pietruszkiewicz, A., and Persson, H. (1963). *Biochim. Biophys. Acta*, **78**, 351–359.

Laurent, T. C., Wasteson, Å., and Öbrink, B. (1969). In *Aging of Connective and Skeletal Tissue* (Engel, A., and Larsson, T., eds.), pp. 68–80, Nordiska, Stockholm.

Laurent, T. C., Preston, B. N., Pertoft, H., Gustafsson, B., and McCabe, M. (1975). *Eur. J. Biochem.*, **53**, 129–136.

Laurent, T. C., Preston, B. N., and Sundelöf, L.-O. (1979). *Nature*, **279**, 60–62.

Laurent, T. C., Sundelöf, L.-O., Wik, K. O., and Wärmegård, B. (1976). *Eur. J. Biochem.*, **68**, 95–102.

Laurent, T. C., Preston, B. N., Comper, W. D., Checkley, G. J., Edsman, K., and Sundelöf, L.-O. (1984). *J. Phys. Chem.*, in press.

Liang, J. N., Chakrabarti, B., Ayotte, L., and Perlin, A. S. (1982). *Carbohydr. Res.*, **106**, 101–109.

Lindahl, U., Höök, M., Bäckström, G., Jacobsson, I., Riesenfeld, J., Malmström, A., Rodén, L., and Feingold, D. S. (1977). *Fed. Proc.*, **36**, 19–24.

Lindahl, U., Bäckström, G., Höök, M. Thunberg, L., Fransson, L.-Å, and Linker, A. (1979). *Proc. Natl. Acad. Sci. USA*, **76**, 3198–3202.

Lohmander, L. S., De Luca, S., Nilsson, B., Hascall, V. C., Caputo, C. B., Kimura, J. H., and Heinegård, D. (1980). *J. Biol. Chem.*, **255**, 6084–6091.

Magdelenat, H., Turq, P., and Chemla, M. (1974). *Biopolymers*, **13**, 1535–1548.

Mandelkern, L., Krigbaum, W. R., Scheraga, H. A., and Flory, P. J. (1952). *J. Chem. Phys.*, **20**, 1392–1397.

Manning, G., (1979). *Acc. Chem. Res.*, **12**, 443–449.

McLennan, H. (1956). *Biochim. Biophys. Acta*, **21**, 472–481.

McLennan, H. (1957). *Biochim. Biophys. Acta*, **24**, 1–8.

Mitchell, D., and Hardingham, T. (1982). *Biochem. J.*, **202**, 249–254.

Mitra, A. K., Raghunathan, S., Sheehan, J. K., and Arnott, S. (1983a). *J. Mol. Biol.*, **169**, 829–859.

Mitra, A. K., Arnott, S., and Sheehan, J. K. (1983b). *J. Mol. Biol.*, **169**, 813–827.

Mitra, A. K., Arnott, S., Isaac, D. H., and Atkins, E. D. T. (1983c). *J. Mol. Biol.*, **169**, 873–901.

Mitsui, Y., Langridge, R., Shortle, B. F., Cantor, C. R., Grant, R. C., Kodama, M., and Wells, R. D. (1970). *Nature*, **228**, 1166–1169.

Morris, E. R., and Sanderson, G. R. (1973). In *New Techniques in Biophysics and Cell Biology* (Pain, R., and Smith, B., eds.), pp. 113–147, Wiley, London.

Morris, E. R., Rees, D. A., Thom, D., and Boyd, J. (1978). *Carbohydr. Res.*, **66**, 145–154.

Morris, E. R., Rees, D. A., and Robinson, G. (1980a). *J. Mol. Biol.*, **138**, 349–362.

Morris, E. R., Rees, D. A., and Welsh, E. J. (1980b). *J. Mol. Biol.*, **138**, 383–400.

Morris, E. R., Powell, D. A., Gidley, M. J., and Rees, D. A. (1982). *J. Mol Biol.*, **155**, 507–516.

Muir, H., and Hardingham, T. E. (1975). *MTP Int. Rev. Sci. Biochem.*, *Ser. One*, **5**, 153–220.

Napier, M. A., and Hadler, N. M. (1978). *Proc. Natl. Acad. Sci. USA*, **75**, 2261–2265.

Nicolis, G., and Prigogine, I. (1977). *Self Organization in Non-Equilibrium Systems*, Wiley, New York.

Nieduszynski, I. A., Gardner, K. H., and Atkins, E. D. T. (1977). *Cellulose Chemistry and Technology*, ACS Symposium Series, **48**, 73–90.

Nieduszynski, I. A., Sheehan, J. K., Phelps, C. F., Hardingham, T. E., and Muir, H. (1980). *Biochem. J.*, **185**, 107–114.

Nishijima, Y., and Oster, G. (1956). *J. Polym. Sci.*, **19**, 337–346.

Norling, B., Glimelius, B., Westermark, B., and Wastesson, Å. (1978). *Biochem. Biophys. Res. Commun.*, **84**, 914–921.

Öbrink, B., and Laurent, T. C. (1974). *Eur. J. Biochem.*, **41**, 83–90.

Oegema, T. R., Hascall, V. C., and Dziewiatkowski, D. D. (1975). *J. Biol. Chem.*, **250**, 6151–6159.

Ogston, A. G. (1958). *Trans. Faraday Soc.*, **54**, 1754–1757.

Ogston, A. G., and Preston, B. N. (1979). *Biochem. J.*, **183**, 1–9.

Ogston, A. G., and Sherman, T. F. (1961). *J. Physiol.*, **156**, 67–74.

Ogston, A. G., and Stanier, J. E. (1950). *Biochem. J.*, **46**, 364–376.

Ogston, A. G., and Stanier, J. E. (1951). *Biochem. J.*, **49**, 585–590.

Ogston, A. G., and Stanier, J. E. (1952). *Biochem. J.*, **52**, 149–156.

Ogston, A. G., Preston, B. N., and Wells, J. D. (1973). *Proc. Roy. Soc. London, A,* **333,** 297–309.

Oike, Y., Kimata, K., Shinomura, T., Nakazawa, K., and Suzuki, S. (1980). *Biochem. J.,* **191,** 193–207.

Pappenheimer, J. R., Renkin, E. M., and Borrero, L. M. (1951). *Amer. J. Physiol.,* **167,** 13–46.

Patrick, C. R. (1961). *Makromol. Chem.,* **43,** 248–250.

Paulsen, H., and Friedmann, M. (1972). *Chem. Ber.,* **105,** 705–717.

Perkins, S. J., Johnson, L. N., Phillips, D. C., and Dwek, R. A. (1977). *Carbohydr. Res.,* **59,** 19–34.

Perlin, A. S. (1975). In *Proceedings of the International Symposium on Marcromolecules* (Mano, E. B., ed.), pp. 337–348, Elsevier, Amsterdam.

Perlin, A. S. (1976a). In *Carbon-13 NMR Spectroscopy of Carbohydrates,* Int. Rev. Science, Org. Chem. Series Two (Hey, D. H., and Aspinall, G. O., eds.), Vol. 7, pp. 1–34, Butterworths, London.

Perlin, A. S. (1976b). *Methods Carbohydr. Chem.,* **7,** 94–100.

Perlin, A. S., and Casu, B. (1982). Spectroscopic Methods, in *The Polysaccharides* (Aspinall, G. O., ed.), Vol. 1, Academic Press, New York, pp. 133–193.

Perlin, A. S., Casu, B., Sanderson, G. R., and Johnson, L. F. (1970). *Canad. J. Chem.,* **48,** 2260–2268.

Perlin, A. S., Casu, B., Sanderson, G. R., and Tse, J. (1972). *Carbohydr. Res.,* **21,** 123.

Phillips, H. O., Marcinkowsky, A. E., Sachs, S. B., and Kraus, K. A. (1977). *J. Phys. Chem.,* **81,** 679–682.

Preston, B. N., and Snowden, J. McK. (1973a). *Proc. Roy. Soc. London, A,* **333,** 311–313.

Preston, B. N., and Snowden, J. McK. (1973b). In *Biology of Fibroblasts* (Kulonen, E., and Pikkarainen, J., eds.), pp. 215–230, Academic Press, New York.

Preston, B. N., and Wik, K. O. (1979). In Wik, K. O., Thesis, University of Uppsala, Uppsala, Sweden.

Preston, B. N., Davies, M., and Ogston, A. G. (1965). *Biochem. J.,* **96,** 449–474.

Preston, B. N., Snowden, J. McK., and Houghton, K. T. (1972). *Biopolymers,* **11,** 1645–1659.

Preston, B. N., Öbrink, B., and Laurent, T. C. (1973). *Eur. J. Biochem.*, **33**, 401–406.

Preston, B. N., Laurent, T. C., Comper, W. D., and Checkley, G. (1980). *Nature*, **287**, 499–503.

Preston, B. N., Comper, W. D., Hughes, A. E., Snook, I., and Van Megen, W. (1982). *J. Chem. Soc., Faraday Trans. I.*, **78**, 1209–1221.

Preston, B. N., Comper, W. D., Checkley, G. J., and Kitchen, R. G. (1984a). *J. Phys. Chem.*, in press.

Preston, B. N., Comper, W. D., Laurent, T. C., Checkley, G. J., and Kitchen, R. G. (1984b). *J. Phys. Chem.*, in press.

Rees, D. A. (1969). *J. Chem. Soc. B*, 217–226.

Rees, D. A. (1972). *Biochem. J.*, **126**, 257–273.

Reihanian, H., Jamieson, A. M., Tang, L. H., and Rosenberg, L. (1979). *Biopolymers*, **18**, 1727–1747.

Rich, A. (1958). *Biochim. Biophys. Acta*, **29**, 502–506.

Robinson, R. A., and Stokes, R. H. (1959). *Electrolyte Solutions*. Butterworth, London.

Rodén, L., Baker, J. R., Cifonelli, J. A., and Mathews, M. B. (1972). *Methods Enzymol.*, **28**, 73–140.

Roots, J., Nyström, B., Sundelöf, L.-O., and Porsch, B. (1979). *Polymer*, **20**, 337–346.

Rosenberg, L., Hellmann, W., and Kleinschmidt, A. K. (1975). *J. Biol. Chem.*, **250**, 1877–1883.

Roughley, P. J., and Mason, R. M. (1976). *Biochem. J.*, **157**, 357–367.

Sajdera, S. W., and Hascall, V. C. (1969a). *J. Biol. Chem.*, **224**, 77–87.

Sajdera, S. W., and Hascall, V. C. (1969b). *J. Biol. Chem.*, **224**, 2384–2396.

Scott, J. E. (1968). *Biochim. Biophys. Acta*, **170**, 471–473.

Scott, J. E., and Heatley, F. (1979). *Biochem. J.*, **181**, 445–449.

Scott, J. E., and Tigwell, M. J. (1975). *Biochem. Soc. Trans.*, **3**, 662–664.

Scott, J. E., and Tigwell, M. J. (1978). *Biochem. J.*, **173**, 103–114.

Scott, J. E., Heatley, F., Moorcroft, D., Olavesen, A., and Casu, B. (1977). *Uppsala J. Med. Sci.*, **82**, 152.

Scott, J. E., Heatley, F., Moorcroft, D., and Olavesen, A. H. (1981). *Biochem. J.*, **199**, 829–832.

Sheehan, J. K., Gardner, K. H., and Atkins, E. D. T. (1977). *J. Mol. Biol.*, **117**, 113–135.

Sheehan, J. K., Nieduszynski, I. A., and Phelps, C. F. (1978). *Biochem. J.*, **171**, 109–114.

Simha, R. (1949). *J. Res. Natl. Bur. Stand.*, **42**, 409–418.

Solursh, M., Hardingham, T. E., Kimura, J. H., and Hascall, V. C. (1980). *Develop. Biol.*, **75**, 121–129.

Stanescu, V., Maroteaux, P., and Sobczak, E. (1973). *Biomedicine*, **19**, 460–463.

Stanescu, V., Maroteaux, P., and Sobczak, E. (1977). *Biochem. J.*, **163**, 103–109.

Stevens, R. L., Ewins, R. J. F., Revell, P. A., and Muir, H. (1979). *Biochem. J.*, **179**, 561–572.

Stivala, S. S., and Ehrlich, J. (1974). *Polymer*, **15**, 197–203.

Stockwell, R. A. (1979). *Biology of Cartilage Cells*, Cambridge University Press, Cambridge.

Stoddart, J. F. (1971). *Stereochemistry of Carbohydrates*, p. 89, Wiley, New York.

Stokes, G. G. (1856). *Cambridge Phil. Soc. Trans.*, **9**, 8–106.

Stokes, R. H., and Weeks, I. A. (1964). *Aust. J. Chem.*, **17**, 304–309.

Stone, A. L. (1965). *Biopolymers*, **3**, 617–624.

Sundelöf, L.-O. (1979). *Berichte Bunsen Ges.*, **83**, 329–342.

Sundelöf, L.-O. (1984). *Anal. Biochem.*, in press.

Tang, L.-H., Rosenberg, L., Reiner, A., and Poole, A. R. (1979). *J. Biol. Chem.*, **254**, 10523–10531.

Thonar, E. J. M. A., and Sweet, M. B. E. (1979). *Biochim. Biophys. Acta*, **584**, 353–357.

Tira, M. E., Calatroni, A., Balduini, C., Torri, G. Moretti, R., and Casu, B. (1979). *Perspect. Inherited Metab. Dis.*, **2**, 165–183.

Tsiganos, C. P., Hardingham, T. E., and Muir, H. (1971). *Biochim. Biophys. Acta.*, **229**, 529–532.

Tsitsishvili, V. G., Grinberg, V. Ya., Fedin, E. I., and Tolstoguzov, V. B. (1979). *Polym. Sci. USSR*, **20**, 2888–2896.

Turner, D. N., and Hallett, F. R. (1976). *Biochim. Biophys. Acta*, **451**, 305–312.

Tyson, J. J. (1976). *Lecture Notes in Biomathematics*, Vol. 10, Springer-Verlag, Berlin.

Vasenin, R. M., and Chalykh, A. Ye. (1966). *Polym. Sci. USSR*, **8**, 2311–2317.

Wang, J. H. (1954). *J. Amer. Chem. Soc.*, **76**, 4755–4763.

Wasteson, Å. (1971). *Biochem. J.*, **122**, 477–485.

Weissberg, S. G., Somha, R., and Rothman, S. (1951). *J. Res. Natl. Bur. Stand.*, **47**, 298–314.

Wells, J. D. (1973a). *Proc. Roy. Soc. London, B,* **183**, 399–413.

Wells, J. D. (1973b). *Biopolymers*, **12**, 223–228.

Welsh, E. J., Rees, D. A., Morris, E. R., and Madden, J. K. (1980). *J. Mol. Biol.*, **138**, 375–382.

Welti, D., Rees, D. A., and Welsh, E. J. (1979). *Eur. J. Biochem.*, **94**, 505–514.

Wik, K. O., and Comper, W. D. (1982). *Biopolymers*, **21**, 583–599.

Winter, W. T., and Arnott, S. (1977). *J. Mol. Biol.*, **117**, 761–784.

Winter, W. T., Smith, P. J. C., and Arnott, S. (1975). *J. Mol. Biol.*, **99**, 219–235.

Winter, W. T., Arnott, S., Isaac, D. H., and Atkins, E. D. T. (1978). *J. Mol. Biol.*, **125**, 1–19.

Yamada, K. M., Akiyama, S. K., and Hayashi, M. (1982). *Biochem. Soc. Trans.*, **9**, 506–508.

Yamakawa, H. (1971). *Modern Theory of Polymer Solutions*, Harper & Row, New York.

Yanagishita, M., Rodbard, D., and Hascall, V. C. (1979). *J. Biol. Chem.*, **254**, 911–920.

# Index